The Bee-friendly Beekeeper

A sustainable approach

David Heaf

The Bee-friendly Beekeeper
A sustainable approach

© David Heaf 2010

2011, Second impression with minor revisions and an addendum

Published in the United Kingdom by
Northern Bee Books,
Scout Bottom Farm,
Mytholmroyd,
West Yorkshire HX7 5JS
Tel: 01422 882751
Fax: 01422 886157
www.GroovyCart.co.uk/beebooks

ISBN - 978-1-904846-60-4

Front cover: View from below of a Warré hive colony just starting its second box.
Back cover: Early April at a Warré hive entrance

Design and artwork, D&P Design and Print
Printed and bound in China by 1010 Printing International Ltd

The materials used in this book are sourced from well-managed forests and certified
in accordance with the rules of the Forest Stewardship Council®

MIX
Paper from
responsible sources
FSC
www.fsc.org   FSC® C016973

The Bee-friendly Beekeeper
A sustainable approach

David Heaf

Northern Bee Books

Contents

PREFACE

This book arose from a series of articles in *The Beekeepers Quarterly* in 2007/8 and resulting suggestions from several people that I republish that material in book form. With the scope that this format has offered, I have revised and greatly extended what appeared in the articles, including, where appropriate, the latest scientific evidence behind some of the views expressed.

Like the articles, this book is intended primarily for beekeepers and would-be beekeepers. A basic knowledge of the life cycle of the honey bee is assumed, as well as some familiarity with elementary beekeeping. This allows me to avoid filling the book with information which is readily available and often thoroughly covered in beekeeping and bee biology publications elsewhere, including the Internet.

Least of all is this book intended to be a manual on beekeeping, although a couple of chapters are devoted to the basics of making and running a simple hive – Émile Warré's 'People's Hive' described in his book *Beekeeping for All* – as one practical example of a possible start on the road to sustainable beekeeping that the author has thoroughly tested.

If the reader requires some advice on where to look for background reading then, for bee biology and behaviour, I suggest Jürgen Tautz's *The Buzz About Bees* (2008). It is superbly illustrated and includes new findings about the natural history of honey bees that may surprise even a few adept beekeepers. Some of these findings are particularly helpful towards designing a bee-appropriate beekeeping.

For a good, recent basic book on beekeeping, which also includes a phenomenological presentation of bee colony development, I recommend Michael Weiler's *Bees and Honey from Flower to Jar* (2006). His book also briefly presents Demeter beekeeping according to biodynamic husbandry, which has a more than 80-year history and involves a way of keeping bees that better respects the nature of the bee.

I refer extensively to material that is available free on the Internet. Some sub-topics on the sustainable beekeeping methods presented here are covered in more illustrated detail on web pages and thus have the advantage that they can be added to relatively easily. The Internet also offers a number of fora and email e-groups which can be very helpful for novices looking for support when starting beekeeping; these help people to keep in touch with new discoveries in apiology as well as to exchange views on keeping bees in a really bee-friendly way.

A selection of Internet resources is included in Appendix 4.

The Bee-friendly Beekeeper

PROLOGUE
TWO COLONIES BECOME FOUR

A bid for freedom?

It is late morning on a sunny day in the middle of spring. A swarm of honey bees has just emerged from its hive in a woodland clearing and is clustered on a thin leafy branch twenty metres from the ground, a few paces to the south of its hive. It sways a little in the light breeze and looks as if it will fall at any moment. The beekeeper arrives and spots it but it is too high for him to reach.

Fig. i: Swarm

There is limited flight activity round the cluster. Some bees are dancing on the outside of it. Others excitedly dive into the middle. Still others have sacs of pollen on their legs. A few larger bees are present – the drones. More minutes pass. Suddenly the swarm explodes in flight. The air is filled with the humming of thousands of bees. In only a few minutes they have disappeared over the trees as a coherent swarm, obviously aiming in a specific direction. The beekeeper sighs. Another swarm is lost. A day earlier and he could have prevented this one too. Entrance activity at the hive soon seems nearly as busy as it was before the swarm left. The bees are completing the process of raising a new queen. The beekeeper sets to work on the routine swarm prevention amongst his hives. Later we shall see what he does.

Meanwhile, let us go back a little in time. For a few weeks scout bees have been making sorties from the parent colony to scour the surroundings for possible new habitations. Many scouts concentrate their attention on a hollow several metres up in a tree that is only a few hundred metres from home. Some fly round the tree trunk seemingly surveying it from all angles. Other surveyors are entering and crawling all round the walls in the dark. The almost circular inside is hardly a foot in width and stretches perhaps twice that length up the tree from a knot hole through the bark near the bottom. The floor of the cavity is covered with debris, almost composted, including fragments of old honey comb from which the wax moths have long since departed. At the dome-shaped top of the hollow are still a few fragments of old comb adhering to the rotting wood.

Each day the number of bees buzzing round the hollow's entrance seems to increase, until, on the day when the swarm first settles near the parent hive, there are sometimes scores of bees present outside the hollow, and the traffic back and forth is busier than ever. Then follows a noticeable lull in activity. But by the time the swarm moves again, enough scout bees have focused on this cavity, one of several found by them, for them to create a quorum among the dancers on the swarm that this is to be the new home for the colony. After the swarm takes off they guide it by flying through the cloud of bees and laying a trail of scent in the air. Within minutes the trunk of the tree round the entrance hole is black with bees and the hole has disappeared. A steady flow of bees, sometimes in layers upon each other, sinks into the tree there. Other bees are standing on the trunk with raised abdomens, exposing a lighter coloured patch at the rear, and fanning their wings. The close observer notices a distinct lemony smell in the air round the entrance to the cavity. The stragglers of the airborne swarm home in on it.

A quarter of an hour passes. All but a handful of bees still fanning have gone inside. For a while it seems fairly quiet round the hole. Inside, the tide of bees has continued to flow as one body slowly up to the top of the cavity. It is as if they are being cautious about what they might encounter. Maybe, but they are also staying in intimate contact with their most valuable member, their queen, without whom they are doomed. On reaching the ceiling, the bees begin to cluster and form an inverted bell shape that hangs down into the cavity, each bee clinging on to the next by linking legs.

In the middle of the cluster they begin to generate warmth, and the temperature starts to rise near to that of the human body. Most of the worker bees have arrived with their crops filled with honey. There is nowhere to put it. Some bees begin to secrete wax scales from under their abdomens and these are used, with much chewing and kneading, to build comb. Others scratch and gnaw away at the top of the cavity, removing old comb and loose rotten wood. A shower of debris and dropped wax scales falls through the cluster to the bottom of the cavity.

By late afternoon, a few bees are already making foraging trips for nectar and plant resins. The resins are being made into propolis not only to reinforce the comb but also to carry on the work of the previous occupants of sealing up the crevices and coating the walls with a sweet-smelling antiseptic film. Within a few hours the first comb has formed, oval and narrow at the top at first, but later heart-shaped. Some of the honey is deposited in the cells near the top. The cells slope slightly downwards towards their ends in the centre of the comb thus helping to keep things in place. Later in the evening the wax makers begin building another comb, parallel to the first and with its centre a little over a thumb's breadth from it. The wax makers now hang in long chains below the combs which are still hidden inside the cluster. The focus of their work is the bottom rim of the comb as it grows downwards. A night of busy activity passes.

Fig. ii: Tree cavity colony. Photo: Clive Hudson

Another sunny morning sees activity round the entrance hole greatly increased. At noon bees are leaving or returning about every second. Some are already bringing in pollen loads of assorted colours, yellows, off-whites, oranges, browns. The queen has laid a patch of eggs in the cells on both sides of the first comb and soon there will be larvae to feed. There is no time to lose, the colony has to establish itself with a satisfactory amount of comb and food stores before winter. On this second day, two new combs have been started, parallel to the others.

At the end of a week most of the width of the top of the hollow is filled with comb, surprisingly orderly and parallel at the top where it is fixed, though somewhat undulating at its lower extremities. The edges of the individual combs are fixed to the walls towards their upper ends and the heart shapes are consequently disappearing. Overall, the combs hang down in a form which almost matches that of the original cluster, only now the edges of the combs are clearly visible and most of the bees who are not out foraging are in the vertical cul-de-sacs between the combs. Guards are already posted at the entrance hole. Other bees stand just inside, fanning their wings. Yet others are busy moulding wax and propolis to restrict the entrance hole to a size more easily guarded against intruders.

The Bee-friendly Beekeeper

The life of the new colony goes on more or less like this all summer. The changes in the flowering plants on offer are marked by changes in the colours of pollen coming in. The complex interaction of the workforce – comprising nurses, warmers, cleaners, disinfecters, food carriers, wax makers, builders, undertakers, honey makers, queen attendants, dancers, foragers – continues amongst the combs. A few hundred new drones are amongst them, helping to keep the brood warm.

Soon most of the combs have a patch of brood. The first brood has hatched long ago and is taking part in the work of the hive. We can picture a spherical or oval shaped zone throughout the nest which is occupied by brood. The comb here is distinctly darker already. Over it and somewhat to the sides is a dome of pollen. Outside that, extending to the sides of the cavity and right up to where the first comb was fixed to the top, the cells are filled with honey, much of it already capped with wax. As brood vacates cells at the upper periphery of the brood zone, the cleaners get to work pulling and chewing out debris and polishing the cell surfaces with propolis. When clean, the cells are filled with honey in the making. The queen would run out of space to lay were it not for the work still in progress to extend comb downwards while the weather is still warm and the nectar flow still plentiful. Although many cells here appear to be unused, some are occasionally filled with incoming nectar, only to be emptied later in the day by workers who move it to where the honey making is in progress. On warm sunny days during the last of the nectar flows, forager traffic at the entrance hole has increased to perhaps two or even more bees returning per second. Many bees leave, never to return as, at the end of their short life of about six weeks, they die outside the cavity.

One small but very important change takes place during the late summer. The bees raise a few new queens in special, larger cells situated in among the combs. When one of these queens emerges from her cell she goes about the nest calling to her unhatched sisters. The tooting and quacking of this conversation is clearly audible at the entrance hole in the quiet of the evening. Soon some of the new queen's attendants scrape wax off the sides of the unhatched cells and she stings her sisters to death. After a few days she briefly leaves the tree on a mating flight. Plentiful drones in the vicinity ensure a good multiple mating and she returns safely to her colony. After a few more days the novice queen works alongside her mother, laying relatively few eggs to start with.

It is now autumn. The days are getting colder and shorter. The nectar supply has almost ended. Comb building has stopped. A heavy store of honey and pollen hangs above the brood nest. The bees have provided well for the long months when the weather curtails both flying and the nectar supply. The brood area has greatly reduced and egg laying will soon end. The bee population is still declining from the several tens of thousands it built up in the late summer to the only ten or fifteen thousand winter bees, most of whom will survive into the spring.

The cluster contracts to occupy the middle of the combs below the food stores. The surrounding combs and the thick walls of the tree trunk give it good protection from extremes of cold. Even on the coldest winter days, the middle of the cluster would feel warm to the hands. Its activity is nevertheless greatly reduced. There is a slow changing of places between bees on the outside and those in the middle as the whole cluster eats its way into its winter fuel supply, the honey dome.

Controlled swarming

Now we return to the beekeeper in his woodland apiary on that day in early summer when he watched his swarm depart. He came there to split some colonies into two in order to increase his stocks and limit natural swarming. He takes the roof off a hive which comprises two brood boxes full of comb and bees. Under the roof is an empty feeder that he has been using to feed a syrup of sugar for the past few weeks to induce early swarm preparations. He removes the feeder and then places the top brood box on a clean floor and stand beside the box that was below it. He begins to lift out and examine the frames of comb of the box he moved, carefully looking at both sides. On the third frame he sees two queen cups. After looking carefully over both sides of the frame he shakes some of the bees off and inverts the comb to look into the cups. They have eggs in. He places the frame beside the hive and continues his search through the rest, seeing more queen cups on other frames. These he destroys. On the sixth frame he sees the queen. He puts her, still on her frame, back into her box and takes a frame from the other box to fill the empty space, making sure that there are no queen cups on it. He moves the queenless brood box, with its floor and stand, to less than a metre to one side of its original position, places in its empty slot the frame with occupied queen cups and places a tank of sugar feed on it before closing up the hive. This colony is losing its foragers, because they return to the old site, so it needs feeding. He puts the queen's hive back in its original position, and places a queen excluder on the brood box followed by a couple of supers or shallow boxes containing empty combs from previous years. Both hives are facing the same way. He puts the cover and roof back on and leaves the apiary. Most of the foragers enter the parent hive with the queen in. They are not only used to returning there but are drawn to the scent of the queen.

One colony has become two, but he has not quite finished yet as one is still queenless. He returns six days later and checks the comb with queen cells in the queenless hive. The cells are near to being capped. He then moves the hive to the other side of the parent hive with the old queen in. Foragers returning from the field

Fig. iii: A National hive colony has been 'shook swarmed' into a Warré hive

are already finding that their home has gone, and after hovering or circling for a while at its old location land on the hive with the queen and soon enter it to unload. The beekeeper does not want a newly hatching queen to leave with a swarm so he ensures that most of the flying bees are with the old queen. He tops up the feeder with syrup.

Nearly three weeks later the beekeeper returns and inspects the frames in the hive that was queenless. The queen cells are empty and torn down. On the fifth frame he finds a large patch of eggs and need look no further. A queen has hatched, been mated and is in lay. Instead of the feeder, he places a queen excluder on the brood box followed by a couple of supers of comb. He has completed one of the dozens of methods of artificial swarming for preventing natural swarming. He induced the swarm process by artificial feeding, but let the bees start raising the new queens in normal swarm queen cups. This time all was under control, although he will watch out for a possible swarm emerging from the parent colony in the middle of the summer.

Balancing interests

The foregoing description of splitting a colony is by no means the most extreme example of the degree of artifice we can apply to raising new queens whilst at the same time making an increase in the number of colonies. But it is nevertheless in stark contrast to the preceding description of a natural swarm. Both processes are subject to bee biology and behaviour except that the first is wholly under the control of the bees. Obviously the first is not beekeeping, although there are beekeepers who keep bees in hollows high up in trees – the Bashkirs of Bashkortostan in the south western Urals for instance, whose techniques of hollowing out cavities have been practised for 1,500 years or more.[1] However, as far as we know this is a relatively uncommon case and hardly feasible or even desirable from the human safety point of view. Even so, in a quest for a sustainable, bee-appropriate beekeeping we can learn a lot from traditional methods, as well as from detailed observation of the natural history of the honey bee, including what is recorded in modern scientific literature. I will thus draw on a wide range of material in what follows.

Fig. iv: Bashkir beekeeper at work. Photo: save-bee.com

The Bee-friendly Beekeeper

INTRODUCTION

Recent decades have seen the appearance of a number of books which promote a beekeeping that is more appropriate for the bee than the kind of beekeeping that is currently widely practised, especially throughout the western world. I am thinking in particular of the books of Matthias Thun,[2] Guenther Mancke/Peter Csarnietzki,[3] Günther Hauk,[4] Phil Chandler[5] and Erik Berrevoets.[6] This book revisits many of the principles addressed by those authors, but sets them in the context of sustainable living and relates them to the 'People's Hive' of Émile Warré more than to any other. Mindful that beekeepers are practical people who, if not actually tired of hearing the term 'sustainability' (the 's-word') are likely to be much more interested in how sustainability can actually be implemented, I have deliberately kept the opening chapter on the fundamentals of sustainability short.

In view of the frequently highly polarised discussions between beekeepers – we are a very opinionated lot! – I have devoted the second chapter to examining a spectrum of possible fundamental attitudes to nature that may be found amongst beekeepers. After all, a bit more self-awareness never did any harm, and might even be found to be helpful in adjusting to the different and perfectly justifiable approaches taken by others of a different persuasion. In this chapter, you can find out if your attitude is that of dominator, steward, partner or participant vis-à-vis the living world and bees in particular.

Bees' needs can be summed up by the three 'S's: shelter, seclusion and sustenance. To survive long in most climates, a honey bee colony needs protection from wind, rain and even from excessive sun. For millennia, beekeepers have provided shelter for their bees in a bewildering variety of ways. I go back to basics in a chapter dealing with thermal and hygrological issues of hive design, always addressing the often conflicting needs of bee and beekeeper. The covering over the nest, the hive body shape, size and composition, and the floor design are the main themes. As the way in which the combs are supported is a fundamental consideration in hive design, this is covered first. We look at a number of hive types, in particular top-bar hives with natural comb, amongst which special attention will be given to the vertical (tiered, storified) top-bar hive of Abbé Émile Warré.[7] Although I give relatively little attention in this book to frame beekeeping as it is generally practised, many of the principles and methods discussed here could be applied by beekeepers who do not wish to relinquish frames. A significant step towards a more sustainable way of frame beekeeping is outlined in Ross Conrad's *Natural Beekeeping*.[8]

Many beekeepers who have learnt their craft in an apicultural milieu which largely favours the convenience of the beekeeper are nevertheless experimenting with relatively natural conditions in the brood nest, in particular with natural comb. As comb is much more than the skeleton of the bee superorganism, the colony, and is thus vitally important, it merits a chapter of its own.

There is also increasing interest amongst beekeepers, especially in northern Europe, to work with locally adapted bees, preferably the indigenous black bee *Apis mellifera mellifera*.[9] Thus it is recognised, at least by some, that after a century, or perhaps longer, of importing bees of other races into a region that is not their home, maybe the local bee is better after all. One possible reason for the preference is that it is likely to be more closely adapted to local conditions after its long evolution in those conditions. By local conditions most people think of the climatic conditions. But could long adaptation to the indigenous forage plants and their seasonality also play a part? We shall look at the whole question of forage, artificial feeding and colony density in the landscape.

If beekeeping in a relatively natural extensive way is healthier for bees, we should be able to identify some principles as to why this is so and hopefully give practical examples as evidence. All forms of husbandry have to cope with diseases and pests and beekeeping is no exception. If we can do it in a way that is healthier, we also improve the economic sustainability of the operation.

According to drawings found in Egypt, honey bees have been semi-domesticated for at least five thousand years. Making increase and selection has no doubt been practised for much of that time. But in the last couple of centuries, breeding has become increasingly artificial. I shall examine what breeding practices sustainable beekeeping could justifiably adopt which really accord with the nature of the bee.

Fig. v: Beekeeping in horizontal earthenware hives in ancient Egypt

There is scope for dealing sustainably with the products of the hive: honey, wax and propolis. I leave out royal jelly production because of the intensification and over-exploitation of colonies it entails. Harvesting and extracting honey by draining and pressing are discussed. Wax recovery and refining as well as propolis collection help increase the economic sustainability of the operation.

A chapter is devoted to the construction and a brief overview of the management of the Warré hive. The hive has seen many modifications, several still in current use. I give a chapter to these and include in it some practical tips from contemporary Warré beekeepers.

1. KEEPING BEES SUSTAINABLY
FUNDAMENTALS

On being asked to write on this subject, I looked around for comprehensive, explicit and concise definitions of the term 'sustainable beekeeping' but was unable to find any that met those criteria. The nearest I got to it was a booklet entitled *Bees and Rural Livelihoods* from Bees for Development, a UK based organisation that works to assist beekeepers in developing countries.[10]

To be sustainable, a human activity should meet the needs of the present without compromising the ability of people elsewhere on the planet, or future generations, to meet their own needs. Meeting needs sustainably is often seen in terms of a 'three-legged stool' whose legs are represented by social needs, economic needs and environmental needs. Take away one of those 'legs' and the whole thing falls down. So to make the particular activity work properly, we have to meet all three kinds of needs.

Sustainability is often seen as a purely environmental or green issue. But, in the context of beekeeping, being sustainable means the activity has to make economic sense in that a beekeeper who lives all or partly from it should have a satisfactory return. And, even for hobby beekeepers, the expenditure has to be proportional to the satisfaction that they derive from it. Beekeeping with the Warré hive, which we discuss in Chapter 9, makes it easier for hobbyists of quite limited means to take up beekeeping.

How does beekeeping meet social needs? Carried out in moderation, and with the proper precautions to avoid it being un-neighbourly or a public nuisance, it can be a healthy outdoor activity which reconnects people with the natural cycle of the year. Its main end product is a beneficial food. Beekeeping provides a pollination service to the community at large, even where no formal hire arrangement is operative. But to meet social needs fully, it should be equitable throughout, i.e. its products should be fairly traded and above all the beekeeper should get a fair share of the overall price paid by the consumer.

Equitableness can be pictured on the global scale by means of the *ecological footprint*. Put simply, it is the area of land that each person requires to sustain their lifestyle. A fair earth share is about 1.8 global hectares (gha) per capita. On average, a person in Britain uses about 5.3 gha per capita. The figure is even higher in some other western countries. It means that if everyone in the world lived like British people they would need three planet Earths to sustain them. The whole thing works only because a majority of people in the developing world survive on far less than one 'fair earth share'. For fairness to be restored, the only logical way that lies within the power of human beings, is for those who live on three planets, so to speak, to reduce their consumption. This option has major implications for beekeeping. *Minimising consumption* is one key criterion of sustainability that we shall frequently refer to in the following discussion.

The ecological footprint can be further divided according to particular material flows. For example, the carbon footprint is part of our ecological footprint that has received special attention in the context of global climate change. How much carbon a beekeeping operation consumes is a factor to take into consideration when judging its sustainability. Before using a particular material we can consider its embodied energy or whether it is from a non-renewable resource. Some operations, for example Royal Hawaiian Honey, go as far as offsetting their carbon consumption by investing in 'carbon-reducing projects such as renewable energy, energy efficiency and reforestation'.[11] However, some regard carbon offsetting as a form of 'greenwash' in that it merely postpones the real solution, namely reducing our consumption to accord with one planet living.

Environmental sustainability also has to take account of the impact on other insects that compete with the honey bee for food resources. Is an area so flooded with honey bees that species diversity is impaired?

As we are concerned with a human activity involving an animal, namely the bee, there is an obvious fourth factor necessary for our activity to be sustainable, which does not seem to fit into the 'legs' of the sustainability stool mentioned above, namely how we treat the bee itself. Do we deal with it in a way that is appropriate to the essential nature of the bee; are our beekeeping methods bee-appropriate, bee-friendly? It is noteworthy that the European Commission regulation on organic production requires observance of the 'species-specific needs' and 'behavioural needs of ... animals'.[12]

In addition to the four factors: society, economy, environment and bee-appropriateness, there is one general and very important overarching principle to be borne in mind throughout this discussion. *What is sustainable in a practical sense varies from place to place on the planet.* For example, we would not recommend making a hive of wood in a region where timber is scarce or where wood digesters such as termites abound.

As the discussion proceeds, it becomes clear that it is difficult to consider just one of the four aforementioned factors in isolation. They affect each other such that we are forced to balance them one with another. For example, it would seem absurd to most beekeepers if they were required to be so bee-friendly that they could no longer rob a colony of its surplus honey. Even so, taking a smaller share of that honey might be more sustainable in the long run.

In saying this I realise that some beekeepers, especially the more commercially minded among professional beekeepers who call themselves *bee farmers*, may regard some or all of what I will propose in the following chapters as ridiculous. In which case, nature will continue to be their teacher. Indeed, we may already be seeing some of nature's lessons in the form of colony collapse disorder (CCD) and the spread of alien pests such as mites and beetles. But in Britain and many other countries, the great majority of hived colonies are managed by amateurs and part-timers working on a small scale. Provided they look after their bees, they are the ones who could have greatest influence on the future well-being of honeybees and the husbandry that goes with them. I hope that that larger group of beekeepers whom I am addressing, those who are not *solely* profit-driven, will find something in this book that will help them make their operation more sustainable.

At this point it may help to look at some fundamental attitudes that we may reasonably expect to find amongst beekeepers.

The Bee-friendly Beekeeper

2. FUNDAMENTAL ATTITUDES OF BEEKEEPERS
ETHICS IN BEEKEEPING

Agricultural and environmental ethics applied to beekeeping

Agricultural and environmental ethics is a relatively new academic discipline. For example, the journal of that name did not start until 1988, and the Food Ethics Council was not founded until ten years later.[13] However, the field is well developed and offers some perspectives applicable to beekeeping.

In my understanding of ethics, an action is only moral in the fullest sense if it is done out of free choice, i.e. with no kind of compulsion from factors such as instinct, norms, culture or religion etc. Furthermore, it is evident that the person doing the action understands and recognises the justification of the moral principle motivating it, that is, that he/she (see footnote) recognises the ideal and is not being forced to follow it. This being so, we cannot prescribe in advance what beekeepers must do. They have to decide themselves on the right course of action.

Disregarding for now the views at the opposite extremes, that of despot and the view that we should not take animal products at all, we can distinguish four possible moral stances or fundamental attitudes *vis-à-vis* nature in general and bees in particular, and support them with ethical argumentation:[14] dominator, steward, partner and participant. The series runs from the most anthropocentric attitude to the most biocentric, or from the most utilitarian position to one of respect for the intrinsic value of the living being. Please note well that no single attitude of the four characterised here is more important or defensible ethically than any other. It is a matter of completely free choice where in the spectrum of attitudes an individual beekeeper places himself.

The dominator

The dominator holds that nature is for supporting the existence of the human race. Therefore it is merely a source of raw materials to serve human goals and to be conquered, controlled, subdued, domesticated. The dominator seeks the maximum utility and profit that is legally, economically and practically possible. Nature on its own follows a course of trial and error by natural selection. But by using breeding technology the trial and error process runs more efficiently, from a

---

1. Hereafter just 'he' will be used, but in all cases 'he' or 'she' is meant.

limited pool of starting material, and at the expense of fewer misfits.

The dominator beekeeper will modify the genetics of his bees by whatever available technique that is also profitable. It may involve artificial insemination and even recombinant DNA technology. If he cannot find the desired genetics locally, he will import them, if necessary from the other side of the world. He will keep his bees in conditions that give him greatest control over them, using frames, embossed foundation, queen excluders, synthetic acaricides, antibiotics, swarm prevention, queen clipping, etc. If possible he will try to breed out any tendency to swarm, and select for supersedure. Mindful of the labour cost, he will open the hive only as often as he needs for maintaining full control. He will take as much honey as possible and leave sugar in exchange. In search of lucrative pollination contracts, he will truck bees thousands of miles. By his supering strategy, including chequerboarding if he deems it profitable, he will make sure that during the main flows the bees perceive empty space above their heads that urgently must be filled.

The utilitarian ethic of the dominator expresses itself in certain practices which, happily, have come to an end or been greatly reduced by more recent hive designs and their management. One is the killing of bees in skeps by asphyxiation over pits of burning sulphur in order to harvest the honey. Another is the formerly routine procedure in certain cold northern climes of killing colonies in autumn to save the costs of feeding and insulating them against the severe winters, and then buying in package bees from the South to repopulate the empty hives. Only the unfavourable economics of this have almost totally put an end to the practice.

The steward

The steward also sees nature from an anthropocentric perspective, but, unlike the dominator, he recognises definite limits. He sees himself as entrusted with the use of nature, not with its consumption. The steward at least endorses a duty to care for organisms other than humans, regardless of the extent to which they resemble humans in their capacity for suffering. The problem is then one of ranking the intrinsic values which the steward recognises as attributes of organisms. Subjecting an animal to a particular form of husbandry or breeding must not happen arbitrarily. Although human interests prevail over those of animals and plants, the latter's interests are more important than purely economic interests. Instrumentalisation of creatures has to be balanced against other considerations. A vivid example of such instrumentalisation is the cow bred to such an extreme for milk production that she needs a 'bra' to support her udder. An example from breeding domestic fowl would be the elimination of broodiness, and from bees, of selecting against swarminess.

Fig. 2.1: Taking breeding too far?

Society accepts that animals can be used for various purposes, but sets limits on that use. However, bees fall outside such control. The steward's duty of care extends to species conservation, protection of ecosystems, to the extent that sometimes human interests must yield to avoid putting nature out of joint. The steward does not want to damage nature's integrity but will gladly domesticate it within limits.

The beekeeper, with the steward attitude to nature, favours the more traditional methods of breeding, but, like the dominator, would resort to modern techniques if a very good case could be made, such as risk of entire loss of the species in a particular region due to an epidemic. He will keep his bees in the way that most beekeepers do, though perhaps avoiding extremes such as queen clipping or not letting the bees winter on their own honey. He will, nevertheless, control swarming in the conventional ways, though try to balance his inspection frequency against the possible harm it does to his colonies. He will also inspect comb for disease and treat accordingly, perhaps favouring non-synthetic acaricides and, instead of antibiotics, comb replacement combined with requeening. He is willing to migrate short distances with his colonies to ensure that they are presented with a good supply of nectar and pollen. He will err on the cautious side when deciding how many hives a particular locality can support.

Bee farmers, a classic example of whom is Robert Manley,[15] tend to occupy the dominator/steward end of the spectrum of fundamental attitudes. Economic sustainability is a powerful influence in determining their husbandry methods.

The partner

The partner regards animals as potential allies, thus presupposing that they have their own 'say' when interacting with humans. He conceives nature as an interplay of life forms, in which each invests its own expressiveness and intrinsic value. This need not conflict with a scientific approach, but does call for a respect for nature. Mankind distinguishes itself from other life forms in that it is not only embedded biologically in nature but also is free to have a conscious relationship with nature, an ethical attitude to it. The partnership is nevertheless asymmetrical, because it consists of the interaction between life forms at different levels of organic complexity. Organic or ecological husbandry satisfies the requirements of the partner but so does a sustainable husbandry that is not necessarily certified organic. In such husbandries, technological exploitation can occur as long as the animal is not unnaturally forced, i.e. its species-specific functions are not prevented. The exploitation might even be of mutual benefit. Biodiversity including diversity of husbanded species is respected.

Compared with the dominator and the steward, the partner beekeeper is willing to accept lower profits in return for maintaining his bees under somewhat more bee-friendly conditions. When breeding bees, he avoids any form of laboratory based genetic modification, though would nevertheless accept conventional queen breeding in mini-nuclei. His hive may or may not contain frames. If it does, he will make them deep enough to contain the brood nest in one box. If he uses foundation, perhaps only as starter strips, it will be from beeswax produced by colonies that have not been treated with synthetic chemicals. If he does not use a queen excluder, he will manage his hive to minimise the chance of the queen laying in the honey boxes. By keeping inspections to a bare minimum, reducing colony density in the landscape, allowing healthy drone populations and letting the bees winter on their own honey he will optimise colony health. If, despite these measures, his bees succumb to disease, he will opt for requeening and comb replacement rather than using chemicals, or he will cull colonies. To control the Varroa mite he will use formic acid, which is already present in the hive or some other natural acaricide that does not contaminate his wax. To raise new queens and make increase he will as far as possible work with the swarming process, intervening to make splits when the time is ripe.

Fig. 2.2: Tree hives. Photo: John Moerschbacher

The partner, and of course the participant, seeks to do his beekeeping in a way that is more respectful of the rest of living nature. Putting hives in certain habitats is a powerful lure to certain wild predators, for example bears in the northern hemisphere and honey badgers in the southern hemisphere. Indeed, beekeepers in Canada are allowed to shoot bears caught in the act. But the beekeeper at the ecocentric end of the ethical spectrum seeks ways of protecting his hives from these creatures, thus avoiding having to kill them. For instance, one Warré hive beekeeper in Canada is experimenting with hoisting hives with pulleys onto difficult-to-reach branches, several metres above ground, to prevent robbing by ground predators such as bears. Another solution is to call in the wildlife authorities to set culvert traps and move the bears away.

Ross Conrad describes an organic beekeeping that espouses the partner approach,[16] although he sometimes refers to it as stewardship.[17]

The participant

The participant sees nature as the totality of interdependent and interwoven life forms. Mankind is an integral part of nature, therefore respect is due to other organisms, not only because of their intrinsic value, but also because of nature's complexity. The innumerable relationships and balances between organisms have a surplus value that exceeds their usefulness to mankind. This has implications that are more far reaching for the participant than for the partner. The participant is more biocentric in his principled choices for setting limits on man's interventions in nature. Although he must also inevitably intervene in nature for the purpose of food production, in doing so he tries as best he can to make use of the inherent dynamism of natural processes. He bases his science and technology on a holistic approach guided by observable phenomena. But participation is not necessarily incompatible with advanced technology. For example, it could be used to investigate the conditions the animal concerned is aiming for, so that husbandry of it can best accord with its essential nature.

Relative to the partner, the participant beekeeper's interests are even more centred on his bees and on their contribution to the natural surroundings. He works with locally adapted bees, raising and selecting them for maximum health. Although he would like to harvest a modest honey crop, he is willing to forego it if there is any risk that by taking it he would have to feed sugar. He is then content to stop at helping maintain a sustainable population of bees in the landscape. He is mindful of the needs of other pollinator species for floral resources and adjusts his number of hives accordingly. He provides 'bee-appropriate' homes for his bees in which relatively natural comb occurs, supported by spales or top-bars. The bees themselves are thus free to determine their optimal cell size and the distribution of cell sizes in the colony, as well as the population of drones required at any particular time. He avoids supering altogether as he sees it placing unnecessary stress on a colony to fill the space that would be constantly appearing above it. His queens are free to roam the whole hive if they wish. He applies the principle that the bees work from the top downwards, so he gives extra space underneath the colony by nadiring. This helps minimise swarms being triggered by lack of space. If the bees are nevertheless intent on swarming, provided that the siting of his apiary is not too urban, he allows them to do so and uses the swarms to start new colonies. He disturbs his colonies as little as possible, maybe even only once a year, and instead observes his bees from outside the hive, learning as much as he can from hive sounds, smells and entrance phenomena. He uses no chemical treatments whatsoever. His Varroa policy is co-adaptation or co-evolution of bee and mite. He harvests his honey, for example by taking one or more boxes of it from the top of his hives provided the boxes are broodless, and he leaves plenty of honey for the colony's winter needs. He has no wish to return to the practice in skep beekeeping of asphyxiating colonies to harvest the honey. He will be reluctant to move his colonies at all unless the natural food supply has become threatened by unforeseen conditions. He welcomes an ethical scientific study of bees, including sophisticated analytical techniques, especially if this will tell him how to keep his bees in an even more bee-friendly way or conserve the species and its habitats.

The attitude of the participant has been linked with biodynamic husbandry,[18] although not all the features of biodynamic beekeeping can be recognised in the foregoing characterisation of the participant. Of the works available in English, Berrevoets' *Wisdom of the Bee* is the most thorough summary of the principles of biodynamic beekeeping.

The ethical matrix

It should be obvious from the foregoing that the boundaries between the different fundamental attitudes are fuzzy. Indeed, an individual beekeeper's attitude to nature and his bees may span two categories or will move between categories depending on the action contemplated and the circumstances. Another form of ethical evaluation is the ethical matrix.[19] Four principles of medical ethics: beneficence (do good), non-maleficence (do no harm), autonomy (freedom/choice) and justice (fairness) are adapted for application to food ethics by merging the first two under the general heading of 'wellbeing' (health/welfare). Four broad stakeholder categories are identified: the animals husbanded (in our case the bees), the producers (beekeepers), the consumers and the biota (the living environment). In this kind of deliberation it becomes a matter of balancing the various interests by considering how the three principles of wellbeing, autonomy and justice apply to the four interest groups. It can be summarised in the 'ethical matrix' (see Table 1).

Respect for:	Wellbeing (Health & Welfare)	Autonomy (Freedom/Choice)	Justice (Fairness)
Bees	Bee health & welfare	Behavioural freedom	Intrinsic nature
Beekeepers (producers)	Adequate income & working conditions	Freedom to adopt or not adopt a particular beekeeping technique	Fair treatment in trade and law
Honey consumers	Availability of pure, safe honey Public acceptability of how the honey is produced	Respect for consumer choice (e.g. organic versus non-organic)	Affordability for disadvantaged groups
The biota	Conservation of the biota (for its own sake and for availability of bee forage)	Maintenance of biodiversity, including bee forage biodiversity	Sustainability of biotic populations.

Table 1: The ethical matrix

The Bee-friendly Beekeeper

In populating the cells of the matrix it is assumed that both consumers and producers generally accept the evaluations to be made. The ones given in Table 1 can of course be modified, removed or supplemented to suit the purposes of the person or persons carrying out the evaluation. The matrix can be used to test any action in the honey production chain to see how it impacts the four interests with respect to the three principles. For example, not everything that the beekeeper does to his colonies has its impact confined to his bees and himself. Using synthetic pyrethroids to kill Varroa results in measurable pyrethroid degradation product concentrations in the honey.[20] This could affect honey consumers. Or, overstocking a region with honey bees could impact the biota negatively by out-competing particular species of wild bee and rendering them locally extinct.

Regardless of his attitude to nature, any beekeeper wishing to undertake an ethical assessment of his own methods may at least find in the ethical matrix a useful way of structuring it. It is presented here with the same four stakeholders given in the source. However, in beekeeping there are two other, albeit smaller, stakeholder groups to take into consideration: the neighbours and other beekeepers in the vicinity. We have to consider if any of our actions impact the wellbeing, the autonomy and the justice that is due to them. This is a very important element of the social aspect of sustainability.

None of the foregoing specifically addresses laws governing beekeeping. All laws originate at sometime in the past from the moral intuitions of men or women. The more glaringly obvious considerations of wellbeing, autonomy and justice are usually the first to end up in law. It has been suggested to me that people with attitudes to nature at the two extremes, namely dominator and participant, might be more inclined to break an apiculture law, only each for different reasons. For example, the profit motivation might lead to compromises with food standards. The presence of antibiotic contamination or even added sucrose in honey would be an example. And beekeepers trying to pursue a totally natural comb policy in jurisdictions where fixed comb is not allowed might be tempted not to register with the authorities and even to hide their hives.

3. SHELTER

... it will be obvious to the intelligent cultivator, that protection against extremes of heat and cold, is a point of the *very first importance*; and yet this is the very point, which, in proportion to its importance, has been most overlooked. We have discarded, and very wisely, the straw hives of our ancestors; but such hives, with all their faults, were comparatively warm in Winter, and cool in Summer. We have undertaken to keep bees, where the cold of Winter, and the heat of Summer are alike intense; and where sudden and severe changes are often fatal to the brood: and yet we blindly persist in expecting success under circumstances in which any marked success is well nigh impossible.

Thus wrote Langstroth,[21] but Pettigrew argued not only that straw was the best hive wall material but also that wooden hives are ruinous to bees because of the condensation that forms on the walls and the mouldy combs that result.[22]

When I started beekeeping, I heard my mentor say from time to time when different ways of doing things were under discussion, 'I don't expect the bees are bothered one way or the other what we do'. In a sense he was right. The honey bee (*Apis mellifera*) seems to be able to make any sort of cavity into its home and even sustains colonies over winter on comb constructed out in the open.[23] However, with the homes we give them and the way in which we manage them, we can take care not to work against the bee's intentions. We can avoid stressing the colony, especially at times of the year when it is less capable of rectifying our mistakes. Coldness/heat, wetness/drought and shortage of food, or insufficient food of adequate quality, if beyond or nearly beyond the capacity of the bee to cope with, can be expected to predispose to stress and thence disease. We will look at the issue of heat and moisture management in the hive first, working from the top of the hive towards the bottom.

Thermal and hygrological issues – 1. Comb support

I deal first of all with how the comb is supported because it is absolutely fundamental to later considerations regarding hive design. The choice of relatively natural comb, instead of frames and foundation, has its own consequences that have to be accommodated in planning the rest of the hive. A more detailed consideration of the properties of natural comb is given a chapter of its own (see Chapter 4).

The Bee-friendly Beekeeper

A bee colony, a super-organism already likened to a mammal,[24] maintains a brood nest temperature of about 35°C, and, even in winter when there is no brood, the temperature is kept as high as 25°C, whereas conditions outside may be many degrees below freezing. The active, targeted, and finely structured nature of this temperature regulation has been revealed by thermal imaging studies in observation hives of individual bees and comb cells.[25,26] The temperature at which young bees are reared affects wing morphology,[27] and the kind of tasks that they take on as adults.[28] Slightly lowered brood temperatures delay the transition to outside hive tasks such as foraging and reduce dance communication behaviour.[29] The behavioural change has been traced to subtle influences on the developing brain.[30] And with temperatures slightly outside the optimum, the workers look normal but show deficiencies in learning and memory.[31] Bees that have been raised at sub-optimal temperatures are more susceptible to tracheal mites (*Acarapis woodi*).[32] Several diseases associated with micro-organism pathogens, for example chalk brood (*Ascosphaera apis*) are more likely to occur if the brood is cooled. Beekeeping that allows the bees to maintain optimum temperatures at all times should help avoid producing bees with such deficiencies in fitness.

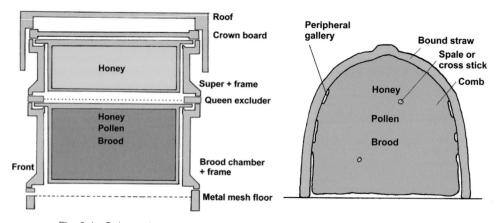

Fig. 3.1: Schematic cross-sections of a frame hive and a skep (not to scale)

The colony is a *warmth organism* par excellence. In nature, the combs are hermetically sealed to the top of a cavity, such as in a hollow tree, and fixed to the walls at the side. This not only provides support, but also retains nest heat. It creates cul-de-sacs of warmed air which, because the warm air rises and has nowhere else to go, helps retain heat in the nest. Renewal of the nest air by diffusion and active fanning by the bees occurs only through the openings between the combs at the bottom of the nest. This natural 'air-conditioning' is under the bees' control. Anything that is done to undermine it happens at the expense of increased activity by the bees. Increased activity necessarily increases the consumption of energy in the form of sugars, the bee's heating fuel, which it normally derives from nectar or honeydew. Whilst skeps perfectly mimic the natural arrangement at the top of a feral colony cavity, modern beekeepers undermine nest heat retention by using frames. These leave air gaps round the sides and the tops of the comb, contrary to the natural situation, creating a bigger surface area on the walls to drive the convection loops which dissipate heat from the cluster. Some beekeepers even winter their bees under a queen excluder with a stack of supers on top. This works only in mild climates provided the sugar supply – often a copious artificial supplement – is adequate to balance the excess heat loss.

Bees abhor empty space. They try to fill in part of the gaps round the frames with wax and/or propolis. Initially this happens on the top-bars under the crown board; but given long enough it also happens between the side-bars and the hive walls. To retain frame mobility, beekeepers must constantly remove the extraneous deposits. Yet as fast as they remove them, the bees replace them. This too increases consumption of food. Thus, on this count, beekeeping using frames, because it increases consumption by the bee and thus by the beekeeping operation, must be less sustainable than not using frames. Furthermore, frames are rarely made by the beekeeper himself but in a factory somewhere. This entails an additional infrastructure, workforce and distribution system. There is much more wood wasted in the manufacture of frames than, for instance, with hive bodies. These factors, taken together, greatly increase the ecological footprint and cost of beekeeping.

Hives which use relatively natural comb such as the skep, the horizontal/long format top-bar hive (hTBH),[33] and the vertical, tiered or storified format top-bar hive of Abbé Émile Warré (vTBH),[34] provided that they are not repeatedly opened, are examples of hives in current use that are inherently more sustainable regarding heat retention.

The horizontal format – sometimes referred to as the Kenyan – top-bar hive comprises a row of top-bars abutting edge to edge along the top of a trough of usually trapezoidal cross-section. The bars form the top of the cavity. The same applies in the Warré hive except that there are spaces between the bars covered with cloth. In both hives the comb is fixed to the top of the cavity and to a large extent to the walls. Compared with framed hives they help reduce energy consumption. Thus they save some stress on the bee, or at least save wasting the bee's efforts on repairing the damage to nest integrity that the movement of frames causes. However, both TBH formats are not necessarily sustainable everywhere on the planet. The horizontal format performs less well in colder climates and the vertical format is not yet thoroughly tested in tropical climates, and is not so easily suited for hanging in trees, which is how hTBHs are often deployed in the tropics.

However, at the time of writing, three Warré hives being tested by Dietrich Vageler in Brazil are performing well.[35]

Fig. 3.2: Horizontal top-bar hive showing comb, top-bars and walls. Photo: Dennis Murrell

For my being made aware of how frame beekeeping can work to the bees' disadvantage, I am indebted to books by Johann Thür[36] and Émile Warré. Thür gives a persuasive argument for observing the principle of retention of nest scent and heat (*Nestduftwärmebindung*) in hive design and resurrects the hive of Pfarrer Johann Ludwig Christ[37] (1739-1813) which was similar in concept to Warré's. In Chapter 10 I discuss a type of frame, originally designed for a Warré hive, that minimises violation of the nest heat retention principle and may offer at least an interim solution in countries where the law requires combs to be very easily removable and replaceable (Delon frame).

Although beekeeping with frames can never be as sustainable in the real sense compared with only top-bars, many of the principles of bee-appropriateness and sustainability that we are discussing could nevertheless be applied using frames. For those who have committed themselves to frames, re-equipping would increase consumption and thus somewhat diminish the resource saving advantage gained. A phased transition would be more appropriate, replacing hives, as they become unserviceable, with top-bar hives. Beekeepers long experienced in the use of frames, which, we have to admit, are there for the convenience of the beekeeper, and who may feel that it is too late to re-equip, should nevertheless find some hints in these pages as to how to make their operation more sustainable.[38]

Fig. 3.3: Warré hive box showing partly constructed natural comb from below

Thermal and hygrological issues – 2. Covering the nest

Roof

What is the most sustainable arrangement at the top of the nest? Heat dissipation here will be potentially higher because warm air rises. The top of the cavity will thus have the greatest temperature difference between the inside and outside. This difference drives greater heat loss in this region. That is one factor to consider. The other is that in hot, or even in temperate climates, the top of the brood nest also needs protecting from solar radiation. A nest cavity in a hollow tree accommodates both these temperature extremes because the leaf canopy shades from the summer sun and the trunk above the hollow contains a considerable thickness of rotting wood that not only insulates but also absorbs moisture.

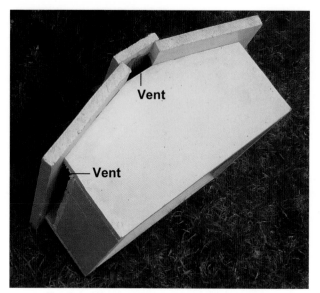

Fig. 3.4: Warré roof showing ventilated cavity

In hot climates, apart from placing hives in the shade, beekeepers try to overcome the problem of thermal control at the top of the hive by including a ventilated cavity between crown board and roof, for example in the National hive in Britain. But such a cavity is far better ventilated if we use a gabled roof having wide-aperture vents at the eaves and gable summit. We refer to this as roof space ventilation. In the Warré hive, this ventilated space does not connect with the nest cavity because inside the roof is a board which rests on the hive and prevents access to the hive by mice and other creatures. It does not have the function of the crown board on National hives. We simply call it the 'cover board' or 'mouse board'.

Quilt

The design of the roof provides for protection from the sun but not for minimising heat loss in the cold. We need something under the roof as insulation. Moving away from dependency on petrochemicals, a non-renewable resource, is a desirable feature of sustainability. Therefore, using synthetic insulating materials in a hive cannot be the most sustainable solution. For insulation, it is possible to devise insulating quilts based on renewables. Indeed, Frank Cheshire's hive has a 'chaff tray' on top of the nest.[39] It is a wooden tray, 75 or 100 mm deep, with loose sacking nailed under it and filled with chaff which beds down over any irregularities. Cheshire writes: 'The loss of numberless colonies is no doubt traceable to defects in the top covering, the non conductive qualities and close fitting of which are far more important than those of the hive side itself'. This chaff tray simulates to some extent the situation at the top of the hollow tree trunk in that there is a thick insulating layer of decomposing wood that can absorb and release water, thus acting as a moisture buffer.

A device identical to Cheshire's chaff tray later reappeared in Warré's hive. He called it the *coussin*, the function of which corresponds to the quilt in the hives of anglophone beekeepers. He recommended using chopped straw, sawdust, wood shavings, leaves etc. as filling. As can be seen from the table in Appendix 1, such fillings have about half the thermal conductivity of softwood, which itself, when dry, is a good insulator. The quilt filling is changed each year, or more frequently if there is any sign of damp. The discarded contents can be used to suppress weeds round the hive.

Whether the moisture buffering property of this unit is significant, remains to be researched. Certainly the material in a Warré quilt is hygroscopic, but there is not yet agreement amongst Warré beekeepers in different climatic zones as to whether the quilt has any significant role to play in moisture handling (see also comment on page 119). However, there is some evidence that the bees themselves control humidity in the nest, although this can be over-ridden by considerations such as temperature control and gaseous exchange.[40]

Villa et al. tested a quilt in overwintering experiments with Langstroth hives as they had heard that beekeepers in the area (Iowa) had used one decades previously because of possible value for humidity control and insulation.[41] It comprised a box with a mesh underneath containing 11 kg. dry oats. They found no significant effect of it on winter consumption of stores.

If the quilt and the roof are made of recycled wood, the hive's ecological footprint will be further reduced.

Nest cover

Most hives in current use have a wooden crown board over the nest. However, as the top-bars suffice as the comb support, we need merely a bee-tight cover resting on the bars to retain nest scent and heat. This can simply be a piece of coarse weave cloth such as hessian – jute sack cloth called burlap in the USA – which is in immediate contact with the top-bars. The exposed surfaces underneath the cloth and the tiny apertures through its weave are soon propolised. It has two major advantages over a wooden crown board. One is that it can be renewed each year at almost no cost, especially if discarded sacking is used. The other is that it can be simply peeled off without annoying the bees. No leverage with a hive tool is required, thus no jarring of the combs.

A further advantage is that a crown board, which is usually made of plywood, is relatively impermeable to gas and water vapour movement compared with a sheet of hessian, albeit propolised hessian. Therefore, this thin fabric, combined with the fact that the gaps between top-bars add up to a much greater area for ventilation than the feed holes in a crown board, allows a greater potential for diffusion.

Cold climates

The top-of-hive configuration described above was developed by temperate climate beekeepers, such as Cheshire and Warré. Its applicability to the far north needs to be considered, for example in the taiga of North America, where the temperature can be far below freezing for weeks at a time. As the top-bar cloth and quilt are permeable to water vapour, condensation could form in it or on the roof's mouse board and drip onto the bees. However experiments with this arrangement in 2008-9 in Alaska and Alberta so far show that there are no reasons not to use it in cold climates too. Furthermore, as the mouse board in the roof is not hermetically sealed to the roof, any water vapour emerging from the quilt here can condense on the vertical surface of the roof's skirt and drain outside the hive.

Thermal and hygrological issues – 3. The hive body

Shape

Fig. 3.5: Weißenseifener Hängekorb. Photo: Ettamarie Peterson

Considered thermally, the ideal shape of the cross-section for a divisible/expandable hive is circular as it presents the smallest external surface to the environment and thus minimises heat loss. The resulting hive would be cylindrical. A hollow tree,[42] a skep,[43] a log hive and many earthenware hives approximate well to this shape. And, as Günther Hauk has pointed out, everything about the bee speaks roundness: hanging and flying swarms; the nest; the queen cell;

the economical packing of round cells resulting in the hexagon shape; the catenary shape of combs; the winter cluster.[44] Matching a hive to the shape of the cluster has been performed meticulously with the Weißenseifen Hängekorb (hanging basket) hive.[45] (See Fig. 3.5)

But roundness is not easily practicable when the material chosen is wood; so the square is the next best choice in the hierarchy of thermally efficient shapes. It is also more practicable for the amateur woodworker. Nevertheless, a number of hobby beekeepers are experimenting with hives that have more than four corners – six, eight and sixteen – thus better approximating to roundness, indeed one of them has developed a divisible round hive out of plaster and straw. These, as yet rare, variants will be discussed in Chapter 10.

Fig. 3.6: Spanish log hive with spales fitted

Fig. 3.7: Modern skep beekeeping in the Netherlands. Photo: Dick van Leeuwen

The Bee-friendly Beekeeper

Materials

The ideal composition of hive walls depends to a certain extent on what renewable materials are available locally. Usually the solution will be wood or some other plant material. Considered just in terms of the materials used, the skep, made of bound straw, has an ideal wall composition because of its insulating and vapour permeability qualities. Indeed, it may even approximate the closest to the conditions of a hollow tree: a domed, cylindrical cavity lined with vapour permeable and insulating material: rotting wood in the case of the tree. The table in Appendix 1 shows that packed straw can have as little as half the thermal conductivity of softwood. However, its measured range of thermal conductivity extends right into that of softwood. The variability is attributed to differences in the type of straw, its moisture content and degree of packing. In a skep it need be packed only to the extent of being bee tight and stopping draughts. And the bees will seal all the cracks with propolis.

There are problems with skeps though. They can engender the barbaric and wasteful practice of sulphuring colonies at harvest time. Their limited rigidity restricts building upwards and thus limits colony size. They are vulnerable to rodents. Also they generally have to be given a secondary shelter, ranging from the simple, transportable and renewable hackle to the resource intensive bee bole.

Matthias Thun overcomes all these problems with his composite straw/wood hive that is robust, extendable/divisible, and which will accept top-bars. Samples of these have lasted over 50 years.[46] His impression is that bees feel better in straw hives. In order to achieve maximum robustness and weathering, the straw used should ideally be rye, as it is high in silica. The outside of his straw hive is cleamed with a clay-dung mixture. Otherwise a secondary housing is needed.

In embodied energy terms, when we consider what it takes to get wood from standing tree to seasoned plank ready to make into hive walls, it seems likely that rye straw has the advantage, even in some of the places where wood is plentiful. But, with modern cereal breeding, it is becoming increasingly difficult to find straw that is sufficiently long. So until straw-hive apiculture is more accessible, wood will be the material of choice for most.

We should also not forget that in some regions the preferred material will continue to be clay. Fired clay hives are relatively easily made in a cylindrical, truncated conical or bell shape and were used not only many millennia ago but also are in current use. They divide mainly into two types. One has the axis of the cylinder horizontal and the other vertical. An interesting example of the latter is Isidoros Tsiminis' top-bar hive whose shape is very closely based on the traditional *vraski* earthenware hives used in Crete until a few decades ago. Although unglazed earthenware is permeable to water vapour and air, i.e. it 'breathes' as do other natural materials, it has a thermal conductivity about ten times higher than softwood. Thus it is more suitable for use in warmer climates. Indeed, that is where clay hives are generally used, for example in the eastern Mediterranean regions and the Middle East.

Fig. 3.8: Isidoros Tsiminis' clay hive[47]

Even if the choice of hive wall material is restricted to wood, it matters a lot which method is used to turn it into a hive. In regions where the appropriate technology is log hives, for example the horizontal format log hives in Jumla, Nepal, wooden hives made from planks are not viewed with favour. They would not only have to be planked laboriously with an adze, or by hiring two sawyers, but also be joined somehow. Straw is not the appropriate choice here either, as it is highly valued as livestock bedding etc. and is prone to rodent damage. Thus the log hive remains the sustainable option and indeed performs well thermologically at high altitude (circa 3,000 m) because of its thick walls.

Fig. 3.9: Healthy colony in side-access log hive, Jumla, Nepal. Photo: Naomi Saville

Partly for reasons of expediency, it has been recommended that people in developing countries should make hives with materials such as metal, plastic and even concrete.[48] Whilst these materials may be more readily available than, for example, wood in particular localities, they do not meet the environmental sustainability criteria of renewability and low embodied energy. In areas where wood, particularly planked timber, is scarce, a first choice would be some other plant material such as rush, reed, straw or cleamed wickerwork.

Wall thickness

We come now to a bee-appropriate design for wooden hive bodies made from planked wood. Water vapour passes through wood, especially if it is not obstructed by certain kinds of paint; and wood gives enough insulation in all but the coldest climates. The desirable thickness depends partly on the expected ambient temperatures. In warm or hot climates about 20 mm would be a minimum, if only for the sake of robustness.

Focusing now on vertical, tiered top-bar hives, such as those of Christ or Warré, a sustainable option for very cold climates would be a thicker wall and, in more extreme cases, possibly a wrapping in winter with some renewables-derived insulation suitably protected from the wet. Wall thicknesses of 38–50 mm have successfully protected wintering colonies in Warré hives in climates where the temperature reaches -30°C, provided that an insulating wrapping is added in winter. In such localities it is recommended that the thermal resistance (R-value) of the walls of any hive be increased to about $1°K·m^2/W$. As a thick wooden wall will give only about half this, the rest can be made up with, for example, fibreboard protected from the rain with a waterproof outer cover.

Box internal footprint

Wall thickness is not the only matter to consider. Another important factor in the thermal performance of a hive box is its internal footprint. A large bee swarm hanging freely is about 30 cm wide. In winter, a Langstroth colony has a cluster diameter of 20-35 cm[49] in a box measuring 37 x 47 cm. If a hive is sized to just fit such a cluster – say 30 x 30 cm – there should be occupancy and therefore warming of most of the width of the inside. Such a size cuts down on the voids that would allow convection currents and helps prevent the condensation and mould commonly associated with the walls and outer combs of wide-bodied hives. With narrower boxes it is possible to use thicker walls without making the extra weight unmanageable. I use 25 mm timber. An advantage of thickening the hive wall is that it tends towards the wall thickness of a hollow in a tree.

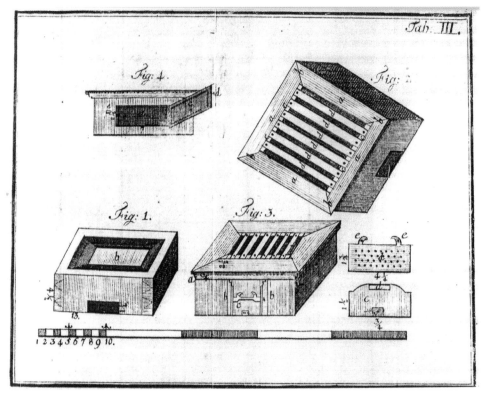

Fig. 3.10: The hive-body box of J. L. Christ's tiered, top-bar magazine hive. Each box, ca. 300 mm square internal plan, has an entrance which can be closed, a shuttered window at the back and six top-bars. The entrance of only the bottom box is kept open.[50]

A narrow inside is bee-friendly for another reason; it places the bulk of the winter fuel supply, if not all of it, above the cluster which naturally works upwards into its stores, and in extreme cold is unable to move sideways. In broad format hives, colonies can starve with peripheral combs of stores only a few centimetres away. Furthermore, warming up a smaller area during the spring expansion uses less energy in the form of honey stores than in a wider format hive.

Depth

How deep should the box be? If the comb is to be removed and replaced, as required by law in some states, it must not be inconveniently deep. But it also must not be so shallow that the brood nest is excessively divided by top-bars. Warré experimented with some 350 hives of several different designs before settling for a depth of 210 mm, which gives a comb depth of about 195 mm allowing for a bee space below it and the thickness of a top-bar.[51] This happens to be only a few millimetres more than the brood comb depth in a National hive. The 210 mm was arrived at partly for convenience in harvesting. The results of his experiments suggested that a box 400 mm deep would be the ideal size for a complete brood nest. But sometimes during harvesting, the bottom of the box still had brood in it, so he opted for boxes half as high.[52] The apparent smallness of a hive element of only 300 x 300 x 210 mm internally is made up for by vertical expansion by adding further elements underneath (nadiring).

Insulation

Whether or not to insulate hive boxes depends partly on the severity of a typical winter. Too much insulation loosens the cluster and thus could increase activity and consumption of stores. Or it could nullify some of the benefit from inso*l*ation, which in the winter falls more horizontally, i.e. on the hive walls. So there is a balance to be struck. Insulating again raises the problem of choice of sustainable insulating materials. One solution is a double-walled hive such as that of William Broughton Carr (WBC) in which a cavity of air between the walls serves as insulation. But this greatly increases complexity, material use, capital cost and inconvenience for the beekeeper. The same applies to Langstroth's double walled hive in which the cavity was filled with plaster of Paris.[53] Furthermore, in experiments with double-walled hives, Warré observed that colonies consumed 2 kg more honey from November to February than those in hives with single walls.

Bee houses have long been used in some countries. This effectively adds a second wall and roof to protect from the weather. However, it is generally a wasteful use of resources. Another traditional solution is surrounding the hive with straw bales protected from precipitation, in which case the obvious question is: why not make the hive of straw in the first place? Furthermore, straw bales prevent winter sunshine falling on the hive walls.

Thermal and hygrological issues – 4. The floor and entrance

The floor must allow adequate ventilation and access during peak foraging as well as the shedding of any condensed moisture. If a solid floor is used, it helps if the inside surface slopes slightly downwards towards the entrance.

Many have found that *open* mesh floors, which have been shown to reduce Varroa in colonies,[54] have also increased ventilation and thus reduced winter mouldering of combs. But other researchers have not been able to show that mesh floors reduce Varroa numbers.[55] So one might ask: does the advantage of expelling a small fraction of the live Varroa justify introducing ventilation conditions that are far outside the control of the bees, especially in winter when they should be conserving energy, and thus stores? It seems not because experiments with overwintering bees show that mesh floors increase consumption of stores by 20%.[56]

Furthermore, as bees appear to control humidity in the nest to some extent,[57] the large aperture to the outside presented by a mesh floor would risk undermining this control. Maintaining high humidity in the nest inhibits Varroa reproduction.[58] The doubtful usefulness of mesh floors in reducing Varroa, combined with the higher complexity, cost and ecological footprint of them, especially of the metal mesh with its high embodied energy, suggests that they are not the sustainable choice either economically or environmentally.

Varroa multiplies faster in a cooler hive,[59] so if mesh floors are used, a drawer under it needs to be closed to retain heat and prevent draughts.[60] This also helps the bees with their thermal control. Given a small but adequate entrance, the bees can create the airflows that they require inside the hive. However, an unfortunate consequence of the drawer being kept in place is that, being inaccessible to the bees, it accumulates hive debris which is a breeding ground for wax moth and micro-organisms, especially when condensation forms on it during nectar flows. Keeping it clean requires extra apiary visits and labour, thereby reducing the sustainability of the overall operation. So on balance, a solid wooden floor is probably the most sustainable solution.

The entrance aperture needs to be sufficient to cope with ventilation require-ments and maximum foraging traffic, whilst not tying up too many bees in defending it or risking 'leakiness' to robbers. To correspond with feral nests thus minimising convection losses the entrance also needs to be at the bottom of the hive.

Warré, who kept bees in northern France, experimented with various entrance sizes and found 120 x 15 mm to be the most satisfactory. This corresponds to an area of 18 cm² which lies in the modal range of 10-20 cm² for feral nests in hollow trees.[61] However, 18 cm² might not be wide enough in regions with more abundant forage, larger colonies and a warmer climate. As with adopting any unfamiliar technology in a new locality, experimentation, observation and reference to local beekeeping experience would be advisable when deciding on entrance size.

The Bee-friendly Beekeeper

4. COMB

Most beekeepers find it impractical to let the bees build entirely natural comb, for example by using, instead of frames, a board at the top of the nest or even top-bars without starter strips. But in the traditional skeps, which have served beekeepers well for over a thousand years, the bees were allowed to make their comb as natural as it would be in a hollow tree. Aside from the constraints of the cavity size and shape, the comb produced in this way has its architecture almost entirely determined by the bees, including comb spacing, cell size distribution and undulations (see figure). In skeps, at most there would be a few thin cross sticks – spales – inserted as support. But now beekeepers usually prefer to guide the bees' comb building to some extent. This can range from plain top-bars to top bars with a beading of wax along their undersides as starter; to top-bars with starter-strips of foundation; to frames with starter-strips of foundation; to frames filled with wired wax foundation or plastic foundation.

Fig. 4.1: Views from below of natural comb in Warré hives and a skep.
The Warré combs are formed without wax starter strips except bottom left.
Photo: Bernhard Heuvel

Drone cells

A full sheet of foundation is the most bee-unfriendly option as it predetermines cell-size throughout the comb. This works against not only drone comb but also worker comb which has its own size distribution in a natural nest. Drone comb is usually regarded as unproductive because hives with it have been found to store less surplus honey, albeit in experiments where it was given artificially.[62] Drones consume stores. So, following Langstroth,[63] drone comb building is discouraged by using worker size foundation and, more recently, for Varroa control purposes, capped drone brood is removed or drone pupae are forked out.

While we do not know the value in the long term of a natural population of drones to the health of colonies, we should be willing to question these practices, especially in view of the facts that natural nests have been found to contain an average of 17% of the comb area as drone comb, and thriving feral colonies to contain in July/August an average population of over 1000 drones (5.6%).[64] It is also reasonable to assume that a beekeeping landscape well populated with drones helps maximise the frequency, genetic diversity and completeness of queen mating. At worst, a poorly mated queen, i.e. one that has been incompletely inseminated, will fail prematurely as an egg layer and, at best, the workers will supersede her by producing a new queen because they can detect when a queen is poorly mated.[65]

Multiply-mated queens produce colonies that are better able to resist the development of brood diseases,[66] including American foulbrood after artificial inoculation with *Paenibacillus larvae* spores.[67] A similar finding was made with inoculation with the fungus spores of *Ascosphaera apis*.[68] Multiple mating confers an adaptive advantage as it increases genetic diversity in a colony and thus reduces the probability that parasites and pathogens will cause a catastrophic infection.[69] Genetically more diverse colonies of honey bees, i.e. those with a higher number of patrilines, are better able to maintain stable brood temperatures,[70,71] and are fitter and more productive.[72] This includes being able to build a new nest more quickly. Genetic diversity is also important for homeostasis generally. Differences between subfamilies in colonies have been detected for the following genetically based task specialisations: preferred forage (pollen, nectar, water or floral location); scouting for food or nest sites; guarding, feeding and tending larvae; grooming and feeding other individuals; fanning in response to high temperatures and undertaking.[73]

As drones contribute to natural selection between competing colonies, i.e. between colonies sending drones to congregation areas,[74] it is important to accommodate a natural production of drones in any management system. Furthermore, it is believed that drones contribute a lot of the food energy they consume to warming the nest. Whilst the importance of their role in thermal regulation is a matter for research, if this is true, it potentially frees up more workers to go foraging.

Worker cells

Letting the bees determine for themselves the size of worker cells may also be important for their health. Worker cell-size of foundation in use in various countries ranges from 5.0 to 5.7 mm,[75] indeed the foundation from my nearest beekeeping equipment supplier is 5.7 mm. The range suggests a certain arbitrariness and disregard for the actual needs of the bee. Furthermore, it is believed that using larger foundation has given rise to bigger bees that now build bigger cells than they used to before the days of foundation.[76] However, this is contradicted by research into old beekeeping literature dating from the 18th century which shows that bees left to their own devices build cells of the same sizes reported in the old literature.[77] Some hold that the practice of using foundation embossed in such a way that it gives rise to cells larger than those in natural comb, has adversely impacted bee health; particularly as regards resisting infestation with *Varroa destructor*. There is a 'small cell' school which holds that bees are better able to tolerate Varroa if the cell size is reduced. This is achieved by using special small-cell foundation to 'regress' the bees to a smaller cell size, usually 4.9 mm.[78] Certainly it has been shown by some that Varroa infestation is lower with smaller cells.[79,80] But more recent work by four independent groups has cast serious doubt on the value of artificially creating small-cell bees.[81,82,83,84]

In any case, regimenting bees with one-size-fits-all foundation is hardly bee-appropriate whatever its cell-size. Measurements of cell sizes in feral or relatively natural nests show a wide variation in sizes from the storage area at the top (5.4 mm) to the lowest part of the brood area at the bottom (4.6 mm).[85] Even in the

Fig. 4.2: Relatively natural comb from a horizontal top-bar hive showing a range of cell sizes.
Photo: Dennis Murrell

brood area, the worker cell size varies. There appears to be no research to indicate the role if any that this differential sizing plays in the functions of the adult bees emerging from the cells.

A way to let bees find their own mix of cell-sizes is to prepare top-bars or frames with no more than a thin starter strip (beading) or lamina of plain wax. The extra wax production in allowing the bees themselves to make almost all the comb, including the midrib, might consume some of the incoming nectar that would otherwise add to the honey surplus, but it is likely to maximise the health of the bee population in colony and landscape. Furthermore, Warré noted in his experiments that bees drawing natural comb took no longer to complete a comb than those given foundation.[86] However, if the instinct to build free comb has become weakened through long use of foundation, wide starter strips of foundation may be needed, at least initially.

Using foundation adds a lot to the complexity, cost and ecological footprint of beekeeping through the milling, wiring and redistribution involved. Furthermore, commercial foundation contains pesticide residues,[87,88] and may contain viable foulbrood spores. The healthy and sustainable option is not to use foundation.

Comb spacing

Without any kind of wax starter, be it bead or lamina, and without any physical comb guides such as 'T' or 'V' shaped top-bars, the bees are entirely free to orientate their comb according to their own requirements in the cavity that they are presented with. However, for many beekeepers this represents a step into chaos, a step too far. It makes the comb unmanageable for the beekeeper. Nevertheless, many beekeepers around the world are using traditional methods, either as a continuation of what has always been done in their area or through re-introduction of such methods. So beekeeping with entirely natural comb, freestyle without even top-bars is obviously far from impracticable. Furthermore, it might bring as yet unknown advantages for the health of the bees.

Certainly there is some arbitrariness about the choice of comb spacing in hives. Measurements of freestyle comb shows that even for European *Apis mellifera* races, or mongrels of them, the spacing is most commonly 30-32 mm between midribs, especially near the centre of the brood nest.[89] Spacings over 40 mm have been seen, especially at the edges of brood nests. Beekeepers using European races space brood combs in the range 35-38 mm. In Warré's hive the spacing between top-bar centres is 36 mm and this is the comb spacing if some kind of guide is used such as a wax starter bead.

If the spacing is left entirely to the bees, then any kind of comb management has to be abandoned. However, in the vertical top-bar hive, such as Warré's, the comb can be handled in units of one box. In Chapter 10 we look at the situation where even top-bars are dispensed with.

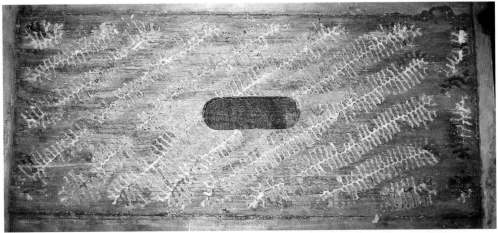

Fig. 4.3: Freestyle comb on the crown board of a nucleus box formerly occupied by *Apis mellifera*. The comb spacing is 30 mm. Photo: Michael Bush

Orientation of comb

It has been long known that bees are sensitive to the earth's magnetic field and can even take their cue for comb orientation from the direction of the field, for example a swarm entering a new cavity 'remembers' and copies the comb orientation of the parent colony.[90] As with comb spacing, we do not know to what extent the subtle influence of the earth's magnetism is needed by the bees. Not giving any artificial guides to orientation would allow the magnetic field to have its full influence. In which case, it would also help to minimise the amount of iron in the hive. It can be present in the form of nails, screws, frame wire, frame rails, queen excluders, floor mesh and roof covering. Some beekeepers use copper or aluminium nails.

Studies of feral nests have shown no overall preferences regarding comb orientation relative to the entrance.[91] This would appear to leave orientation up to the beekeeper if some form of comb guides are being used.

Comb age

In feral/wild colonies there is a natural turnover of comb. Unused young comb from a previous season that has not been reinforced with cocoons can be all or partly torn down and remodelled. Furthermore, if a colony dies or eventually abandons a cavity, many scavengers including wax moths and rodents can gradually reduce the comb to powder. Thus, in nature, there is a varying degree of comb renewal.

The reasons for including comb renewal in hive management are that old comb is thought to accumulate disease causing organisms and the cells have been shown to have a slightly reduced internal diameter due to the build up of cocoon material. Research has indeed indicated that in dark, heavy, old comb, compared with new comb constructed in the same season, the cells are slightly smaller and the resulting bees weigh slightly less. But, paradoxically, old comb showed a small increase in brood survival in week two, although the area of sealed brood was 20% more with the new comb.[92] However, the overall hive population at the end of the season was not significantly changed by the type of comb. So there is no unequivocal message from this study regarding comb renewal and we can conclude that renewing comb routinely is not actually harmful. Furthermore, opinion appears to differ among beekeepers on whether queens prefer to lay in new comb or old although in the study cited here the authors concluded that the queen lays preferentially in new comb. Some beekeepers with long practical experience have seen no harmful effect of using tough old comb provided it is structurally sound and even regard it as better for wintering bees.[93]

Frame beekeepers generally aim to renew comb to some extent each year by replacing two or three old combs with frames of foundation, thus establishing approximately a five-year cycle of renewal. In more extreme situations, especially if European foulbrood has taken hold, all the comb is changed, for example by shook swarming and destroying the old comb. For changing just a few combs, the foundation is often inserted in the brood nest in order to promote rapid construction, rather than at the side, where the foundation could easily go stale before the bees work it. The disadvantage of this method is the splitting of the brood nest which is disruptive to its integrity and heat retention. A Bailey comb change is a more organic and thorough way of comb renewal while retaining the brood nest until new comb is available.

In horizontal top-bar hives there is no generally agreed system for comb renewal but it can be achieved relatively easily by placing empty top-bars at the side of the nest, suitably primed with a wax starter or a few millimetres of comb.

In vertical top-bar hives (vTBH), comb renewal is built into the process of harvesting from the top of the hive and nadiring new boxes, i.e. placing empty boxes underneath, and thus avoids all disruption of brood nest integrity. As brood vacates cells at the top of the nest, they are preferentially filled with honey and eventually the whole brood nest moves down into lower boxes. This is how it takes place in a natural bee nest, leaving honey in the upper regions.[94] In the course of a season a colony in a vertical TBH achieves the same as a Bailey comb change.

Comb renewal also automatically takes place with any hive where the comb is cut out at harvest time, for example skeps and log hives.

Fig. 4.4: Renewal by bees of dry comb from the previous year in a Warré hive

Role of natural comb in colony communication

The comb is more than just the skeleton of the bee superorganism, it also provides a communication system both chemically and mechanically. The role of pheromones in ordering colony life has long been known, and it is not unreasonable to propose that the pheromonal communication system is highly structured three dimensionally throughout the nest. For example, if the 'dance floor', the patch of comb on which the communication dances are performed, is moved to a new location in the nest, the dancers move to the new location. Another example is brood pheromone which influences both the physiology and behaviour of other members of the hive. If there is indeed a spatial organisation of pheromones then the practice of arranging combs for the convenience of the beekeeper may be more stressful physiologically than is merely removing and replacing combs in their original order. The integrity of the three dimensional pheromone structure of colonies may be a clue to what Johann Thür meant by the retention of nest scent in his term *Nestduftwärmebindung*.

Beekeepers are generally aware of the importance of nest integrity. The routine inspection of combs followed by replacing them not only in the right order but the right way round, is common practice. And there is also an awareness of the more subtle aspects of the integrity of the three dimensional structure, albeit not yet extending to 'pheromone structure'.[95,96]

The comb is also probably important for mechanical transmission of sound communication between bees. Jürgen Tautz describes it as a 'telephone network'.[97] If comb is fixed in frames, its freedom to oscillate and thus transmit signals, is diminished. In the region of the 'dance floor', the bees chew away the comb or foundation to leave a wide gap between the comb and the frame. Clearly, in hives with freely suspended natural comb this extra work for the bees would be unnecessary. We might also question the effect of foundation, which generally creates comb with thicker midribs than that of natural comb, even though the foundation midrib is thinned by the bees during comb construction.[98]

5. SECLUSION

Disturbance unsettles colonies. Probably the commonest form of it is the intentional disturbance by the beekeeper himself. Every opening of the hive that lets the heat out forces the bees to repair the damage – repairing broken comb and repropolising – and to restore the 'thermal structure' of the colony by extra heat-production activity. In winter, the cluster can take as much as three days to return to normal.[99] Depending on the extent of comb manipulation, even in summer the restoration of the pre-opening condition could take as much as a day. Brood that has been chilled in spring provides an opportunity for foulbrood to get a hold.

The repair work after intrusion is done at the expense of other activities, and certainly increases energy consumption, thereby reducing stores and/or the honey surplus. Almost all the incoming nectar that is available to the bees is used in heating the nest.[100]

Opening a colony that is already coping with the challenges of pests and disease may tip the balance towards its succumbing to them. Thus, to be bee-friendly, such interventions should be minimised whilst nevertheless maintaining good management. This could involve just one hive opening in the real sense per year, namely at harvest. A vertical top-bar hive, such as that of Warré, makes possible this low frequency of intrusion because new elements are added only below the brood nest, i.e. by nadiring. In nature the brood nest grows sideways and downwards, and combs can extend downwards to as much as 1.5 metres. Adding boxes below, allows the colony to expand indefinitely and does not let the heat out of the brood nest because the nest can be lifted intact, together with its covering. A simple manual lift allows a single operator to do this without any obvious disturbance to the bees.[101] Full boxes of honey are removed from the top, if possible only once, namely at the end of the main nectar flow.

Even a horizontal top-bar hive can be worked, to some extent, without letting the heat out of the brood nest. It is worked from the back, i.e. away from the entrance, and the top-bars, which abut one another, stop the nest heat from dissipating by convection of the warm air of the nest. Combs can then be harvested or moved further back and new space given.

The extremely infrequent disturbance possible with the vertical top-bar hive can even be maintained in areas where there is a legal requirement that bees must construct comb in frames which can be removed for inspection. It would simply be a matter of arranging inspection to take place during the annual opening of the top of the hive at harvest time, preferably when there is still brood to be seen. Even beekeepers with the commoner frame hives could consider how to adjust

Fig. 5.1: Feral colony in cavity wall.
Photo: Joe Waggle

their management to make intrusion negligible.

Moveable frames are a relatively recent phenomenon in the history of beekeeping. Could they be a contributor to the rise of bee diseases and epidemics in modern times? Such a question could be answered by a well conducted, ideally multi-regional, research programme comparing frame and fixed-comb hives or non-intrusive versus intensive husbandry.

Controlling swarming is another reason that colonies are disturbed at frequent intervals, sometimes as often as weekly, and over periods of up to five months with some beekeepers. But really bee-appropriate management should accommodate not oppose the bee's instincts. Arguably the most inconvenient of those instincts in management terms is swarming, and beekeepers use all sorts of methods to reduce it. Certainly swarming and the way we deal with it can impact all three aspects of the sustainability of an operation: economic, environmental and social. For example, we often have to double-up on equipment at least for a while. But just letting swarms escape represents an economic loss to the beekeeper and could even cause a public nuisance. So can we justifiably extend a policy of non-intervention to swarming? We shall look at some options for this.

The most time-consuming, but spectacular, form of letting swarming happen that I have come across was used in Georg Klindworth's Lüneberg skep beekeeping operation in Germany, at least until the 1970s. The beekeepers wait for first signs of swarm issue and catch them in nets.[102] This will suit only the most dedicated beekeepers with endless time to spare. An alternative to this, which does not require the beekeeper to be present, is a swarm trap over the hive entrance. But even traps have the disadvantage of being relatively labour intensive and require a fairly accurate knowledge of when a swarm is due, especially if they are not to result in the death of many drones. Another option is to place bait hives with swarm lures in suitable sites near the apiary. Details of these are given in Chapter 8.

Fig. 5.2: Bait hive of recycled tongue-and-groove timber prominently sited on the roof of a garden shed

One trigger of swarming is lack of space in the brood nest. This can be overcome by giving plenty of space ahead of demand. This was long ago observed by skep beekeepers who put ekes (straw rings) or nadirs (hive bodies) underneath heavy colonies so as to give more space for the colony to grow downwards.[103] This is especially easily done by the bottom-expansion method of vertical top-bar hives described above, ensuring that at least one element is present below the growing brood nest. No weekly comb inspection is needed. Indeed, such inspection might even provoke swarming. We shall return to the matter of swarming in the context of breeding in Chapter 8.

An empty element at the bottom of the colony also allows disturbance of the colony during artificial feeding to be almost eliminated. A container of feed can be placed on the floor. No heat is let out of the brood nest. Colonies appear rarely to notice the intrusion.

Fig. 5.3: Feeders on Warré hive floor, and viewing colony development with a mirror.
Photo: John Haverson

Non-intervention surveillance

Beekeepers accustomed to opening hives frequently will probably find the bee-friendly, non-interventionist approach advocated here difficult to adjust to. Admittedly, after even my relatively few years of frame beekeeping experience, not opening the hive can be a little frustrating at times. So what are the alternatives for surveillance?

Firstly, consider the history of the colony: the date of hiving and weight; its expansion rate; whether it has swarmed etc. Secondly, there are the conditions in the environment to take note of: has the weather been so bad soon after hiving that the bees could not forage and are likely to need emergency feeding? Thirdly, observe the activity at the hive entrance. Is the colony foraging at the same rate as others in the apiary? If foraging is feeble, depending on the colony's history, it could be failing.

Does the proportion of foragers returning with pollen look about right? If not, there may not be a laying queen. Would you expect there to be one already or was it not so long ago that the colony swarmed? A lot of information can be gained from entrance activity. Is it purposeful or are the bees scurrying around as if looking

Fig. 5.4: At the hive entrance ('Der Bienenfreund' (1863) by Hans Thoma,
with the kind permission of Staatliche Kunsthalle, Karlsruhe)

for something? Storch's excellent book on the hive entrance phenomena is worth reading for more advice on this.[104]

Fourthly, we can heft the hive to check the status of stores. This can be done by just lifting one side a fraction, or by using a simple weighing device (see Chapter 9). Fifthly, listen to the colony by putting an ear to the box. Can you hear the normal bustle of a good sized colony or is it very faint. Are queens tooting and quacking at one another indicating queen replacement is in progress? If you have cause for concern about queen status compare the sound after a gentle tap with a knuckle. Is it a 'hiss' that's gone in a second or a prolonged sound likened to a moan, a wail or even grumbling? If you have difficulty hearing the cluster then push a listening tube a little way into the entrance. It greatly amplifies the sounds. I use a cheap stethoscope with the diaphragm removed. Sixthly, feel immediately above the brood nest, the top-bars on a horizontal hive or the bottom of the quilt filling in a Warré hive. Is it as warm as you would expect? And finally, does the draught of air from the entrance on a sunny day in the warmer months have a wonderful aroma or an unpleasant odour. If the latter, an advanced stage of foulbrood is likely present.

Novices, while gaining confidence in beekeeping, may like to incorporate a shuttered window in each hive body box. Horizontal top-bar hives are sometimes fitted with windows and the practice of having them in each element of vertical top-bar hives goes back more than two hundred years.[105] A modern example is in the Frères-Guillaume version of the Warré hive (see Chapter 10).[106] Adding windows is less sustainable though, because of the high embodied energy of glass and the increased complexity of construction. Also, because of the relatively high thermal conductivity of glass (see Appendix 1), it requires good insulation behind the window shutter, usually polystyrene, which brings with it the added disadvantage of being a non-renewable material.

There is a simple alternative to windows for just keeping an eye on how far comb building has progressed: in the vertical hive it suffices to bore an inspection hole of about 40 mm diameter in each box through which a small lamp can be shone. This hole is plugged when not in use. Information can also be gained from the tray under a mesh floor, if used. The debris can be examined, if necessary with a hand lens, to discover evidence of what is going on above: Varroa mites including ones that have been damaged; cappings; shedding old pollen; larval and pupal fragments; wholesale dismantling of old comb; pollen loads etc.

Fig. 5.5: View through Warré hive window

Inspecting for disease

The most prominent criticism of the kind of frameless, low-intervention beekeeping that is a feature of vertical top-bar hives, such as the Warré hive, is that it makes inspecting for disease more difficult. It is argued that it is almost a reversion to skep beekeeping.

In the United Kingdom, statutes have existed since 1942 to allow official inspection of colonies for foulbrood. For example, with American foulbrood, the period since 1942 has seen a general fall in its incidence. It dropped from about 1.5% of managed colonies known to the Inspectorate in the period 1952-1954 to 0.6% in the period 2002-2004.[107] This is believed by proponents of state-funded foulbrood control to be a result of the institution of an inspection service.

The Bee Diseases and Pests Control Orders 2006 now provides not only for the control of foulbrood but also the compulsory notification of the disease, i.e. beekeepers must tell the authorities if they suspect that they have an infected colony. This places low-interference beekeepers in something of a dilemma. If foulbrood must be controlled by opening hives and removing combs for inspection, something that disrupts colony life and heat retention, and causes stress, setting it back a day or more, then how are beekeepers to comply with the Control Order whilst continuing to manage their colonies in the most bee-appropriate and arguably the healthiest way?

It is worth looking at what the inspection service is achieving. The incidence was low to start with and was reduced by only a little over half. The pathogen *Paenibacillus larvae* is ubiquitous in 'sub-clinical' amounts,[108,109] and where not detected in samples is likely to be present in numbers below the limit of detection of the bacteriological or DNA amplification techniques used. Therefore, it would not be unreasonable to conclude that the vast majority of bee colonies are coping with the disease by means of their natural immunity and hygiene processes, at the levels of both the individual bees and the colony as a whole. Given that foulbrood eradication is completely out of the question, for this incidence to remain low and maintain the current level of co-adaptation, some exposure of bees to foulbrood is necessary. Without that, there is no selective pressure.

Averaged over ten years, the UK Inspectorate inspects at least 25,000 colonies each year.[110] This is only about 10% of known managed bee colonies.[111] There is also an additional unknown number of hived and feral/wild colonies. From the figures we can deduce that, on average, a managed colony has a chance of being inspected once in every ten years. On the one hand this is good news for the low intervention beekeeper as the chance of having official disruption and possibly damage to a natural comb colony is not very high. But on the other hand it places a high degree of responsibility on the individual beekeeper to monitor for foulbrood, either for signs of a failing colony as discussed, or when removing boxes that have had brood in.

In 2009, the official surveillance of hives in the UK was increased somewhat to 15%, presumably because of the sharp rise in colony mortalities for various reasons.[112] Even so, this level is still only a small proportion of known colonies. The authorities are no doubt balancing the existing incidence of foulbrood against the availability of resources and are clearly not operating a foulbrood eradication programme.

In addition to the low level of official surveillance, we should consider the manner in which inspection is conducted by opening colonies and studying the individual cells of brood combs. Colonies found to be infected are destroyed. This approach is like locking the stable door after the horse has bolted.

With modern bacteriological and DNA amplification methods, the pathogen can be detected at appropriate tolerance thresholds *before* a colony has succumbed. Pathogen counts on/in bees in certain parts of the brood nest are good correlates of the presence of foulbrood. This approach would also have the advantage of reducing the extent of intrusion in colonies which itself could promote brood chilling and disease. It would no longer be necessary to inspect individual combs. Instead, a sample of bees could be taken from the honey area of the colony and submitted for analysis of *Paenibacillus larvae* colony forming units. Testing bees from this part of the colony has been found to correlate 100% with disease incidence.[113] And with both the horizontal top-bar hive and the Warré hive, the sampling would occur without significant loss of heat from the brood nest. With a Warré hive it would be sufficient to peel back the corner of the top-bar cloth and sample bees from the honey box as the bees come up to investigate the light and what is letting the heat out.

If, however, the local Inspectorate is inclined to treat natural comb hives unsympathetically, then the option is always available to make them fully inspectable without completely compromising their underlying principles. With the horizontal top-bar hive this would merely mean keeping comb management ahead of the bees' tendency to create cross-combs. With vertical top-bar hives such as the Warré hive one can use half-frames, Delon frames, or just manage the comb so that it can be removed with a top-bar comb knife such as the one designed by Bill Wood. These options are discussed in Chapter 10.

With the growing interest in using 'hives for bees' rather than 'hives for beekeepers' – to quote two of Frank Cheshire's chapter titles[114] – disease inspection policy and methods will need to be revised sooner or later to avoid damage to healthy colonies living on natural comb.

Unintended disturbances

We should also consider disturbances by agents other than the beekeeper. This will only be brief as the manuals on beekeeping usually cover this particular nuisance quite well. Intrusion by people other than beekeepers, namely vandalism and theft, of course reduce an operation's economic sustainability. If we can, we keep our hives well screened and away from thoroughfares. Roger Delon, who kept over 600 modified Warré hives in the Vosges/Jura region, found that he needed to padlock and chain them under communal steel roofs in fours with concrete blocks set in the ground as anchors.[115] The extra materials and cost of this were clearly necessary to make beekeeping viable in that locality, but it seems hardly sustainable in the full sense. Such precautions are not unreasonable though. The first Warré hive apiary in Germany was destroyed, apparently by beekeepers.

The risk of major disturbance, even destruction, of colonies by animals varies a lot according to locality, and therefore local knowledge amongst beekeepers would be the first point of reference. Beekeepers have always sought to protect their hives from such disturbance, if only for the sake of economic sustainability. Protection from farm livestock is generally relatively easy. But the list of intruders that present more of a challenge is almost endless: insects (wasps, small hive beetle), rodents (mice), birds (woodpeckers), mammals (bears). Solutions include insect lures, fencing off apiaries, wire netting the hive, putting it on a platform accessible only by ladder, hanging it in trees or keeping it in a building. A simple skunk cage is shown in Fig. 10.5. The more elaborate the defences, the less sustainable the operation and there may come a point where the extra materials used make its continuation at a given site hard to justify on environmental, let alone economic, grounds. Interestingly, the logo of *Bees for Development* is a Kenyan top-bar hive in a tree.[116] What could be simpler?

Finally, extraneous vibration is a source of disturbance worth taking into consideration. Passing trucks or trains shake the ground. Even the drumming of heavy rain or hail on flat, metal-clad roofs has been identified as a cause of colony disturbance. Such roofs are more likely to transmit this vibration to the colony than sloping roofs made only of wood. If the drumming happens in winter, it could loosen the cluster and risk increasing consumption of stores.

African bees and absconding

Whilst the vertical type of top-bar hive has so far been tried in only a limited way with Africanised bees, it seems to perform satisfactorily, at least in Brazil. What is of more interest in the context of allowing bees seclusion is whether it works for African bees, i.e. *Apis mellifera adansonii*, which are very prone to absconding if the colony is interfered with. There are no reports yet of trials of vertical top-bar hives in Africa but one might predict that the low intervention beekeeping made possible with the vertical top-bar hive would reduce the probability of colonies absconding. Firstly, extra space is added from below, thus leaving the brood nest undisturbed during the process. Secondly, when harvesting honey, it is done when the brood nest and most of the bees have moved down out of the top box(es). However, just as complications arise if horizontal top-bar hives are not hung horizontally when sited in trees, even worse complications could arise if vertical top-bar hives are not hung vertically.

6. SUSTENANCE

Forage

Beekeepers are already aware that they should put their hives where there is enough nectar (and/or honeydew) and pollen within flight range throughout the flying season. In one study, the median foraging range was found to be 1.7 km and 95% of foraging was within 6 km.[117] Foraging of up to 14 km has even been reported,[118] but the benefit/cost ratio decreases as the distance increases. Surveying from time to time what forage is available, and what the bees are visiting within a kilometre or two radius, seems prudent because what is discovered could be used to help decide on stocking density.

Food availability over a whole foraging season largely depends on floral diversity and the consequent more or less wide range of flowering times. But, in our cultural landscapes, floral diversity and quantity are severely declining, helped by the use of herbicides and the over-manicuring of marginal land and hedgerows. This impacts honey bees as much as it does other species that depend on flowering plants. In my own locality, dominated by grazing, my bees could have abundant clover. Certainly the fields are sown with clover mixes, but local farmers, in their war against dock, often manage to eradicate their clover at the same time. Furthermore, commercial varieties of white clover sown seem not as attractive to honey bees as the wild type.

Obviously the bees need a sufficient quantity of nectar and pollen, in the latter case it may be as much as 25 kg. per season.[119] Less obvious for maximal bee health is the requirement for *diversity* of pollen. Indeed diversity of pollen has been found to increase immunocompetence in honey bees.[120] Pollens vary greatly in quality, especially as regards the amino acid composition of their major nutrient, protein. Pollens differ in their capacities to supply natural antagonists of bacteria/fungi and antibiotic substances,[121] such as certain fatty acids.[122] Lack of pollen diversity decreases a colony's antimicrobial defence faculties and it becomes more susceptible to disease. A variety of pollens allows one pollen to balance the shortcomings of another, for example too low a protein or fat content or too high a mineral content.[123] Indeed, honeybees are generalists in their pollen gathering, evening out and diversifying it.[124]

Compared with pollen diversity, nectar diversity may not be so important qualitatively, but a variety of melliferous plants is of value for maintaining a steady supply of nectar throughout the year. Also, bees prefer nectar, if it is available, to digging into their honey stores.

It is usually not practicable for beekeepers to compensate significantly for poor floral diversity and/or abundance by planting, as they usually lack control of sufficient land to make an appreciable difference. However, we can have a certain amount of influence, such as by planting suitable herbaceous plants, shrubs and trees; lobbying for agriculture that has regard to biodiversity, and only purchasing products from such an agriculture. We could encourage neighbouring landowners to plant appropriately, for example when we give/sell them a sample of the honey we produce. A neighbour of one of my apiaries delights in the profusion of honey bees that visit his ornamental heathers and poached-egg plants. While the plight of the honey bee is so frequently in the media, the time is ripe to raise awareness of bee plants.

A more subtle consideration regarding forage is whether bees in a region are adapted to the forage in that region, as was suggested by Rudolf Steiner in a series of lectures he gave in 1923.[125] I cannot adduce any observations in support of it, but it seems worth considering. I can envisage that bees may be adapted to plant toxins or antinutrients of one region but not of another. And many beekeepers recognise the advantages of working with locally adapted bees. For example, the Bee Improvement and Bee Breeders' Association (BIBBA) seeks to identify and preserve the indigenous bee in Britain and Ireland.[126] Here, as amongst beekeepers generally, what is in mind is adaptation to the climate and seasonal availability of forage, rather than the regional flora itself. If adaptation to the flora is important, then moving bees that are adapted to one region to another region, with an annual spectrum of forage that they are not accustomed to, could pose an added stress. Both buying colonies from far away and long distance migratory beekeeping, particularly the kind practised by beekeepers providing pollination services, are examples of what could give rise to such stress.

Colony density in the landscape

How many hives should a single apiary have? Too many in relation to floral resources risks stressing the bees and thus compromising their health. Competition with feral colonies will exacerbate this. Maximum densities of ferals in an area with no managed colonies have been found to range from 7 to as high as 12 colonies per km^2, although estimated pollen and nectar resources would have supported far higher numbers.[127] But those low numbers may result from another factor determining feral density, namely minimising spread of disease or infestation due to drifting. As low as one colony per square kilometre has been found among forest-dwelling wild honey bees that have co-adapted with Varroa in the USA.[128] In contrast, in Australia, which is so far free of Varroa, a study found a range of 0.1 to 77 feral colonies per square kilometre in landscapes ranging from heathland to riparian forest.[129] The same study pointed out that with a foraging range of 2 kilometres, 100 hives in a single apiary amounts to less than 10 colonies per square kilometre. Using genetic marker techniques, colony densities in Europe were found to be 5-6 colonies per square kilometre.[130] This corresponds to either 72 colonies in a single apiary, with no others within a 2 kilometre radius, or individual hives dotted over the landscape at 400 metre intervals.

It has also been recognised that natural control of the tracheal mite (*Acarapis woodi*) depends on good foraging opportunities and reducing competition between colonies, i.e. on a low stocking density. Infestation of the mite coincides with high local colony densities.[131]

Probably only a few beekeepers would accept a stocking density as low as one colony per square kilometre (0.01 per hectare) especially if it means a kilometre between individual hives. So a compromise will usually be made. The density of my colonies, which are situated in four apiaries fairly close to one another, is 1.6 per square kilometre, taking the median foraging radius as 1.7 km, and allowing for overlap of the foraging range by that of another beekeeper's colonies about 2 km away.

Fig. 6.1: Garden apiary of three Warré hives

Judging exactly how many hives a site should hold will usually be a somewhat subjective procedure involving weighing up forage quantity and quality as well as competition from known colonies, managed and feral, nearby. In parts of Australia, stocking densities are under official management. But without such organised guidance, the experience of previous beekeepers at a particular site will no doubt help in deciding. The aim is to avoid over intensification. This is because disease rates are directly proportional to stocking density, as in any branch of husbandry.

Let us encourage extensive, i.e. low-intensity, beekeeping carried out by a larger population of beekeepers and made possible by much simpler equipment that requires far less intervention. Top-bar hives, especially the Warré hive with the minimal intervention it requires, make an increase in small-scale beekeeping a realistic possibility.

Fig. 6.2: A load of Warré boxes and a hive lift has just arrived at the apiary ready for the spring nadiring

In order to have more hives without over-stocking, beekeepers have more or less distant fixed apiaries or they practise transhumance. They may do this partly to avoid having to feed bees artificially because a particular locality becomes devoid of nectar at a particular time whereas another, perhaps with a specific floral source such as heather is still available. Other reasons include providing pollination services, obtaining particular floral honeys and boosting overall yields. Whatever the reasons, it entails the use of motor vehicles and the consumption of fossil fuels to power them. Indeed, migratory beekeeping, on the scale it is practised today, is largely the result of the availability of motor vehicles and relatively cheap fuel. But for the beekeeper who wishes to work sustainably, there is the difficulty of deciding where the balance lies between fossil fuel consumption and avoiding artificial feeding. It applies whether beekeepers travel between their fixed apiaries or move their bees between temporary apiaries near seasonal floral sources.

Out of personal choice, I keep my apiaries within reasonable cycling distance, and ferry equipment around with a bicycle trailer (Fig. 6.2). For me, having to drive my car to my bees would detract from the enjoyment of the hobby.

An advantage of the low-intervention beekeeping that is possible with the Warré hive is that it may if required be visited only twice a year. This could help towards minimising the impact of travel on the sustainability of a particular beekeeping enterprise.

Feeding

Should a beekeeper, who is aiming to be bee-friendly and sustainable, feed his bees? Ideally, no. We generally expect the bees to feed us, not we them. So routine feeding is not the ideal option. However, there are bound to be emergencies when a colony, through no fault of its own, i.e. through no inherent genetic weakness or failing queen, falls on hard times. It may be because of a long spell of bad weather or through some manipulation that the beekeeper has done, such as artificial swarming to make an increase. These are the times when we can justifiably feed.

But the need for routine feeding, say in preparation for winter, can almost always be eliminated by leaving sufficient honey for the bees. The beekeeper takes only the surplus, which in some years might be none at all. This seeming sacrifice of some of the harvest, and therefore of profit, can be minimised by using a hive in which colonies winter very economically. Warré, beekeeping in northern France in the first half of the 20th century, consistently found that his 'People's Hive' needed 12 kg stores for winter, whereas his Dadant hives needed 18 kg.[132] This difference has since been confirmed by others.[133] The better thermal performance of the Warré hive largely accounts for the difference. Indeed, I have had Warré hives winter on 4 kg stores and build up to swarm strength the following year.

What to feed to the bees is a source of endless controversy amongst beekeepers. The sustainable option is not to feed sugar, least of all high fructose corn syrup. They are usually products of intensive monocultural agriculture and are processed in a chemical refinery before being transported long distances. The energy consumption, environmental degradation and pollution involved do not justify their use. Routine sugar feeding entails its procurement, additional equipment, preparation, distribution and cleaning – all of which add complexity, cost and labour to the operation. If sugar must be used, then the gold standard would be organically certified, refined, i.e. crystallised, sugar. The choice of the actual plant source for sugar, whether cane or beet etc., depends on its relative environmental, social and economic impact. Food miles, supporting sustainable livelihoods, and fair trade should all come into consideration.

That feeding sugar could have consequences for the bee's ability to combat disease is indicated by experiments on populations of lactic acid bacteria in the bee larva gut.[134] The gut contains a spectrum of such bacteria which can totally inhibit the growth of *Paenibacillus larvae*, the pathogen of American foulbrood. Feeding sugar syrup has been found to reduce lactic acid bacteria counts in the bee gut.[135,136] Perhaps there are other subtle consequences of feeding sugar which might not be outweighed by its advantage as a quick supply of fuel to the bees.

If sugar is not fed, then it leaves only honey as the possible primary energy source. Honey is the natural choice. A sustainable beekeeping operation retains sufficient comb or extracted honey from its own apiaries to satisfy emergency feeding. Furthermore, life cycle analysis has shown that, compared with sugar, honey is the sustainable and ethical sweetener.[137] A compromise would be to feed a sugar and honey mixture. According to Demeter (biodynamic) beekeeping guidelines, chamomile tea and a trace of salt should be also added.[138]

But two objections immediately arise regarding feeding honey to bees. One is that the honey used, even honey from the same apiary, may contain sufficient foulbrood spores to spread infection. This objection is based on the 'germ theory of disease', i.e. that micro-organisms *cause* disease. In the next chapter I argue that while they are a necessary condition for disease they are not a sufficient cause. Even so, obviously one wise precaution would be not to feed honey from a colony or apiary known to have had foulbrood or even Nosema. The micro-organism counts in the honey may be way beyond the capacity of the recipient colony's defences, especially as it is already most likely under stress through having to be fed artificially.

The other objection to feeding honey is that, over long northern winters, sugar gives better colony survival than honey, especially where honey stores have a relatively high pH, ash, conductivity and/or protein content, for example honey from honeydew or heather. But it really depends on how well the bee is adapted to its local forage. For example, a less thrifty bee imported from the south will quite likely require its winter food to be imported from the south too in the form of sugar. In sharp contrast, in north America, north of 50 degrees, colonies survive on honey stores for six months or more as the nectar flow season is very short.

Similar arguments to the above apply to the choice between feeding pollen or pollen substitutes. If pollen diversity and quantity are low, the question must be asked as to whether the site is at all suitable for sustainable beekeeping.

If a beekeeping operation can be managed so as to rely on nectar and pollen as sole nutritional inputs, then it can be supported by any combinations of marginal land, gardens, woodland and crops that need pollination. No additional land needs allocating to produce food for bees, least of all land for sugar production monocultures.

7. DISEASES AND PESTS

We have so far discussed the factors that we can optimise to create healthy living conditions for bees. They are appropriate shelter; natural comb, including its renewal; seclusion; adequate sustenance of the right quality and colony density. Despite our best efforts to make conditions for our bees as ideal as possible, there is an additional factor that often calls for our intervention if our bees are to thrive. This is the destructive pressure from diseases and pests. On the one hand we can regard them as selective pressures that keep the race of bees we are using in good shape, but on the other hand there are more or less frequent circumstances where colonies fail to cope and some kind of action becomes essential. In this chapter, rather than cover all the possible diseases and pests, which can easily fill more than a third of practical beekeeping books, even those on natural beekeeping,[139] I pick out a few of the more serious threats to use as examples of a more sustainable approach. I also include a short section on pesticides which, for over seventy years,[140] have impacted beekeeping, especially in areas of arable farming and horticulture.

Varroa

Among pests and diseases the *Varroa destructor* mite has the biggest economic impact, more than foulbrood or Nosema, at least in the UK. As discussed in the previous chapter, managing Varroa by maintaining extremely low colony densities will not appeal to many. Instead, beekeepers try to make conditions in the hive unfavourable to Varroa. One way they do this is by using chemicals. These increase in acceptability to the organically inclined in the sequence: synthetic pyrethroids, thymol, organic acids and powdered sugar.

There are two primary reasons against putting synthetics into hives. The first is the burden on bee health through the direct effects of the chemical or through their having to detoxify or otherwise deal with it in their metabolic processes. Researchers have found it difficult to source foundation wax that does not contain acaricides.[141,142] Of the chemicals put in hives, those used to control Varroa take up the greatest portion by weight. All the foundation manufacturers whom I contacted on this matter say they do not monitor wax residues. Toxicology recognises the potential health burden from small traces of substances migrating from manufactured products including from plastics and from finishes. By 'out gassing' or dissolution, these extractables present a challenge, usually concealed, to living systems. We should try to keep them away from our hives and wherever possible use materials to which organisms have had a much longer time to adapt, namely natural materials. To avoid chemical introduction via foundation, it should be home made or from a clean source, for example Demeter (organic) certified. Such wax commands a higher price as it is preferred by cosmetics manufacturers for its low residues content.[143]

The other primary reason is the impact on human health through ingesting chemicals via the honey produced, for example 3-phenoxybenzaldehyde, a degradation product of the acaricide tau-fluvalinate.[144] Other reasons include the associated increase in the complexity of the operation and labour involved; the increase in its ecological footprint through these chemicals having to be manufactured, a process that usually produces a far greater weight of waste than the weight of product produced; and the increase in cost of production of honey which is especially high if patented acaricide delivery systems are used.

To what extent can we do without acaricides? One approach, namely using foundation with a reduced cell-size, was mentioned in Chapter 4. But in bee-friendly beekeeping we do not use foundation. Dusting the bees with powdered sugar is commonly used. However, it has been found to be ineffective by one group who dusted with 120 g sugar every two weeks for 11 months.[145] Perhaps the frequency of dusting was too low, but even at the frequency used it represents a severe intrusion into colonies, as it means repeated opening of the brood nest, something to be avoided if possible for reasons already discussed.

Some colonies in frame hives started with conventional foundation are surviving for between seven and ten years with no chemical treatment.[146,147,148] In a study over 17 years, swarming and survival rates of colonies initially dropped after Varroa arrived but eventually recovered to their original values.[149] Added to this, there is an increase in the number of anecdotal reports from beekeepers, some with several hundred colonies, in regions where Varroa has been present for a couple of decades, claiming that colonies, especially those under minimal intervention management, are surviving without treatment. Also, feral/wild honey bee populations are recovering or have recovered in some areas.[150] This has been attributed to a reduction in Varroa virulence as the bee and the mite co-evolve[151] as well as natural selection for resistance in bees and introgression of selected resistant bee genetic material.[152] Whatever the reason, it is encouraging for those who would like to avoid treating for Varroa.

Another factor suggested for feral recoveries is interruption of the brood cycles by natural swarming. This gives a clue to a non-chemical Varroa management strategy. French commercial Warré beekeepers have found that artificial swarming gives satisfactory control.[153] At peak foraging time in warm settled weather, all the bees are swarmed into a new hive. The brood is left on the site of the old hive and the new one is taken to another apiary out of flying range. Returning field bees repopulate the brood combs and raise a new queen. Egg laying and therefore Varroa reproduction is interrupted in both halves. The snag with this method is that a method of reproduction is used which is not the most bee appropriate – the artificial swarm combined with the creation of emergency queens. We will discuss this in more detail in the context of breeding in the next chapter.

Whether breeding Varroa tolerant strains is viable *vis-à-vis* the degree of monitoring and breed-purity maintenance involved remains to be seen.[154]

Certainly in the long term, to step off the treadmill of Varroa treatments, mite and bee will need to co-adapt and this means tolerating a mite population that is sufficient for mite and bee fully to interact with each other. Certain beekeepers in several countries, including the occasional commercial beekeeper,[155] have colonies that are well along the road to co-adaptation, but their colony losses would be unacceptable to some. Even so, I believe it is the only sustainable solution to the Varroa problem.

Fig. 7.1: Natural comb in an octagonal Warré-type hive box. Photo: Dietrich Vageler

But if chemicals have to be used just to keep colonies alive while co-adaptation is taking place, for example in the case of beekeepers with only one or two hives for whom a loss would be a severe blow, then treatment needs to be part of integrated pest management, i.e. monitoring Varroa burdens and treating only when absolutely necessary, and then at the most appropriate time. Some beekeepers get by with treating only every two years. The organic acids are preferable for the low level of residues that they leave behind in colonies. Essential oils, such as thymol, are less acceptable because of their absorption into wax. However, we must not lose sight of the fact that any intervention on behalf of the bee or against the mite just further postpones the achievement of co-adaptation.

European beekeeping has not yet had to deal with the challenge of small hive beetle (SHB). It appears that strong colonies, with no crannies to hide in, will cope with the SHB challenge. Top-bar hives with their natural comb may prove well suited to this.

Micro-organisms

We discussed Varroa first because it has the biggest economic impact for beekeeping, at least in Europe. Next in severity come foulbrood and dysentery, the latter commonly involving *Nosema apis* or *cerana*. Both are associated with micro-organisms that we describe as 'pathogens', implying that they are the cause of the disease. This is true for a particular point of view. In contrast there is the view that if, through our husbandry, we create the conditions for disease, for example by over intensification, frequent hive opening and other stresses, then the micro-organisms, which are generally ubiquitous in small amounts,[156] will find conditions in which to proliferate. That *Paenibacillus larvae* (American foulbrood) spores are generally found throughout the honey bee population, especially in managed colonies, has been shown by studies in several countries.[157] Given the right conditions, the micro-organisms will start to proliferate. They become a symptom of the disease process. Whereas beforehand, only ultra-sensitive detection methods would detect them, now there are visible changes in the vitality of the colony.

The inhibitory action of lactic acid bacteria on *P. larvae*, is just one example of the role of beneficial micro-organisms in bee colonies. The full story is likely to be very complex. Healthy bees and their food have been found to contain thousands of micro-organisms including bacteria, moulds and yeasts.[158] They have co-evolved with the bee over millennia and their presence in the right balance is likely to be compromised by the introduction of acaricides and antibiotics into the colony, for example organic acids and thymol. Whilst the toxicity of these chemicals to bees has certainly been studied, their more subtle, indirect affects, via their impact on the beneficial microbial population, has received less attention.

With the completion of the sequencing of the honey bee genome it has emerged that the bee has distinctly lower immune pathway redundancy and thus flexibility compared with other insects.[159] However, this is compensated for by social or group level activities for coping with disease. The spreading of propolis throughout the hive, including on comb and in cells, is an example of a behaviour that helps create an environmental barrier to disease. Propolis has been found to reduce the expression of immune function related genes and lower bacterial loads in colonies.[160] There may also be other species-specific behaviours that are part of the 'social immunity' of the colony.

The hive atmosphere has been found to contain formic acid and acetic acid.[161] Both of these substances are known inhibitors of the growth of micro-organisms. Could they be part of the social immunity of a hive that is also contributed to by propolis? If they are this would lend further support to Johann Thür's hypothesis of *Nestduftwärmebindung* – the retention of nest scent and heat – which he envisages as best preserved by allowing natural comb to be built without the use of frames.

So that bees can cope with challenges from micro-organisms by using their own defences, sustainable beekeeping tries to create bee-appropriate conditions in hives. At the beginning of this chapter were listed a number of obvious management areas that could be optimised. One measure not mentioned there is the role of natural swarming in disease control. Foulbrood can be regarded as nature's way of weeding out weak stocks, and American foulbrood is an exception amongst bee diseases in that it is particularly virulent. Under natural conditions, vertical transmission of pathogens via swarms is the most important route of infection of new colonies.[162] Theory predicts that vertical transmission should generally select for relatively benign pathogens. However, beekeeping practices reinforce each other in enhancing disease transmission. They favour horizontal transmission of pathogens through high colony densities, provoking robbing and moving equipment and bees between colonies (splits). At the same time, beekeepers control swarming, thus removing an important mechanism for the evolution of low virulence, namely vertical transmission. So it would not then be an exaggeration to say that foulbrood is a problem of the beekeeper's own making. This may partly account for why foulbrood incidence is higher in managed colonies compared with ferals.[163]

If, despite our efforts, disease visibly manifests, then we cull the hopeless cases; artificial (shook) swarm into a clean hive those cases that merit it, for example colonies with European foul brood (*Melissococcus plutonius*); and, if it is a minor ailment, let the colony remedy the matter itself. In the extreme case of culling and burning, the economic loss with a top-bar hive is a lot lower than with one filled with frames and foundation. Above all, we do not keep diseased colonies that we try to prop up with antibiotics.

As the foulbroods are particularly virulent diseases, beekeepers should of course be vigilant for signs of them and submit their colonies to official inspection where required. We have discussed ways of attenuating the impact of this in Chapter 5.

Pesticides

Books on beekeeping published even as late as the 1940s do not mention the adverse influence of agricultural and horticultural pesticides on bees. However, with the increasing use of agrochemicals since the Second World War, the problem has become steadily worse. In recent years, the severest cases have been the large-scale *acute* poisonings of bees on the European continent by neonicotinoids. But it is recognised that the *chronic* toxicity burden of pesticides is one of the factors contributing to the high losses associated with colony collapse disorder in commercial operations, especially in the USA.[164] While having to cope with subacute poisoning by a pesticide or cocktail of pesticides, the bee's organism is under stress. When an additional potential stress is present, such as a pathogen, then the chance of succumbing to disease is higher. Evidence for such interactions is emerging in the case of the neonicotinoid insecticide imidacloprid and the microsporidian pathogen Nosema. A recent paper shows that the two agents work synergistically to impair bee health. The activity of the enzyme glucose oxidase, which allows bees to sterilise brood food, is reduced compared with controls when both agents are present.[165]

Most beekeepers will not be able to site their apiaries sufficiently far from land where pesticides are used, especially given that foraging bees can travel as much as 14 km from their hive.[166] The other options are to establish good communication channels with local farmers who grow crops known to be treated with pesticides so that early warnings can be obtained, continue to lobby for better controls on pesticide type and use as regards toxicity to honey bees and boycott food that is produced with the use of pesticides by buying certified organic products. The long term cumulative effect of this economic pressure may eventually result in a reduction in pesticide use.

8. BREEDING AND MAKING INCREASE

As well as the economic value aspect of sustainable beekeeping, namely harvesting honey and other hive products, we should not forget the need to balance the impact of beekeeping on the environment against its more general benefit to society through the provision of pollination and as an enjoyable pastime. Indeed, it is interesting to note a recent intensification of interest in the benefit to society of beekeeping, not only in the media but also from the 2008 UK government survey of beekeepers. Among other things this was interested in finding out how many beekeepers would give up replacing colonies that they had lost to diseases and pests.[167]

That there have been larger losses than usual in many countries in recent years is beyond dispute. For example the average colony survival in USA over the 2007/8 winter was estimated at 64%, slightly higher in the UK. All sorts of reasons have been put forward including aspects of management, for example the practice of trucking thousands of colonies from state to state, which cannot possibly be regarded as either bee-friendly or environmentally sustainable. But a central aspect of management has rarely come under scrutiny, namely breeding practices. In Chapter 4, in relation to having plenty of genetically diverse drones and thus the need to avoid drone-suppression measures, I discussed the importance of genetic diversity for bee colony homeostasis and health. Here we examine how queens are raised and colony numbers increased.

Artificial queen breeding

In a lecture series on bees in 1923, Rudolf Steiner warned about the problems that artificial queen breeding would introduce eventually, prophesying that beekeeping would be in danger in 80 or 100 years time.[168] A beekeeper who was contributing to the lecture series queried this and was told that the effects of modern breeding would not show up at first. Steiner was referring to something more subtle, namely the forces in the hive that had hitherto been organic were becoming 'mechanised'. He said that the intimate relationship between a colony and its queen that has been raised naturally cannot be achieved with a queen brought in from elsewhere. Breeders and their customers used to profitable honey crops from their queens will no doubt think this statement absurd. But might there be something in it? The aforementioned 64% winter survival rate is only as high as it is because colonies are propped up by medication and by feeding imported sugar. Take away those props and what would the survival rate be? Also, that rate is an average. Some commercial concerns lost 90% or more.[169] Steiner was not alone in warning that artificial queen breeding produced inferior colonies. His near contemporary Émile Warré, a beekeeper with personal experience of artificial breeding, wrote: 'artificial breeding that is practised in intensive beekeeping only gives mediocre and inferior queens. Again, the stock will not benefit from this. As a result people will end up only with bees that are weak, poor workers, incapable of resisting disease, above all foulbrood.'[170]

Whilst the queen breeding aspect should certainly not be overplayed, it seems worth keeping it in mind as a factor in bee health, especially since considerable public money is being diverted into trying to discover why colonies are dying out. So let us look at it in some detail and at the 'mechanisation' of breeding to which Steiner referred. Mechanisation increases in a series which progresses from queens raised in the natural process of swarming; to splits made where no swarming impulse has started; to modern queen breeding including grafting larvae etc; to artificial insemination; and ultimately to recombinant DNA technology, which has been suggested as a way of creating queens capable of producing colonies with disease resistant traits.

There is a distinct difference between a normal queen whose larva is raised from hatching in copious amounts of royal jelly in a specially prepared round, domed cup hanging vertically, and an emergency queen whose larva is initially raised in an almost horizontal hexagonal cell from an egg or larva that had been destined to be a worker. Beekeepers have long known that colonies often supersede such emergency queens with normal ones. The two types of queen may look the same but the bees can tell the difference. All commercial queens sold as queens are emergency queens, only they are usually raised in round cells from the outset after grafting the larva into a plastic cup. Alternatively, they may be started by confining the queen in a plastic cage mounted on a comb in the colony so that she is presented only with an array of plastic cell 'plugs' in which to lay. The plugs are then transferred to plastic holders that maintain them in the normal vertical orientation

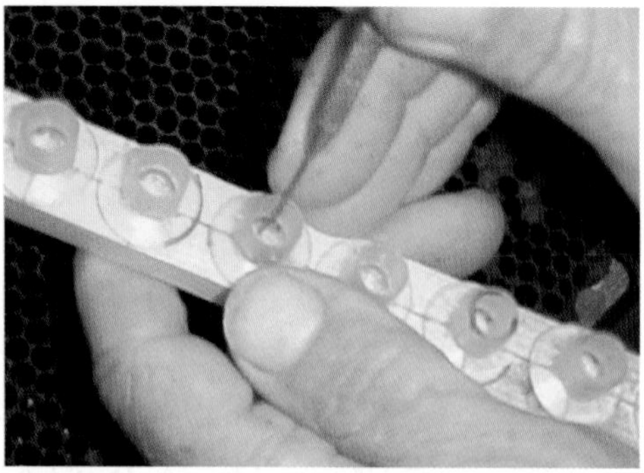

Fig. 8.1 Grafting larvae.
Photo: Tom Glenn, glenn-apiaries.com

in which queen cells are constructed and placed in a queenless colony.

In order to introduce into a colony a queen resulting from this process, the colony needs to be brought to the point of 'desperation' to receive her. And once the queen is introduced, she is liable to be replaced by one that the colony has produced itself, one in which the 'intimate relationship' that Steiner mentioned has a chance properly to unfold.

Fig. 8.2 Queen cell frame and close-up. Photo: Tom Glenn, glenn-apiaries.com

The physical basis for the intimate relationship between queen and colony may by highly complex, but one obvious factor is queen substance. Research into queen substance is probably still in its early days, but it has long been clear that it is closely connected with colony health and organisation. It controls the behaviour of workers, for example reproduction, foraging, raising new queens and swarming. The multiplicity of its effects are partly context related and partly due to its complex composition. Indeed, queen substance, in particular queen mandibular or retinue pheromone, has been found to comprise nine chemicals and there is evidence that it contains others as yet unidentified.[171] As 170 odour receptors have been detected in honey bees, it seems not improbable that the chemistry of the intimate relationship between queen and colony will turn out to be even more intricate. Work continues in several laboratories to unravel it right down to the molecular level of receptor chemistry and associated gene expression. Important for our considerations here is the finding that 'the production of queen pheromone is exquisitely sensitive to factors associated with reproduction and mating'.[172] In comparing naturally mated queens with those that were artificially inseminated, it was found that the former exhibited the greatest ovary activation and the most distinct mandibular pheromone chemical profile.[173] Multiply mated queens are more attractive to workers.[174] This attraction is part of the intimate relationship between them.

In the light of these recent findings, we can justifiably ask the question: is the receptor-pheromone chemistry and the resulting behavioural responses of workers something we can instantly 'switch on' by introducing a queen to a colony or does it take time for the various classes of worker to adjust their behaviours to the spectrum of pheromones available? If the latter, then the longer-term exposure of the colony to the maturing queen may be essential for full development of the intimacy of the relationship. This means in-colony breeding, not raising queens in mating nucs at a commercial breeding station and airmailing them around the globe.

Artificial queen breeding has been identified as one of the most damaging contributors to loss of genetic diversity of honey bees worldwide.[175] For example, in the USA most commercial hives have progeny from as few as 500 breeder queens. Added to that there has been a tendency for beekeepers and breeders to favour a limited choice of subspecies, for example in Europe *A. m. carnica, A. m. ligustica* and *A. m. caucasica* out of the ten available there. As with all husbandry, genetic uniformity in a given area exposes the beekeeper to catastrophic losses if disease strikes.

A few other aspects of artificial queen breeding deserve mention in this context. Purchased queens are rarely from the same locality as the one in which they are used and are thus not locally adapted strains. In husbandry there was a great wisdom in the development of landraces, i.e. breeds adapted to the locality in which they were raised. And recognition of the value of local adaptation partly motivates efforts to conserve the European black bee, *Apis mellifera mellifera*, in its original habitats. The race evolved to cope with the climate and forage of the region without needing the prop of imported sugar, and was/is no doubt rich in locally adapted sub-types. Using a locally adapted bee is the sustainable option, especially as it does not require a whole industrial breeding infrastructure and the carbon-inefficient distribution system that goes with it. But if that is the case, then the whole task of breeding falls on the beekeeper, or at least to very local co-operatives of beekeepers. It is to be welcomed that the Bee Improvement and Bee Breeders' Association (BIBBA) aims to work with local groups of beekeepers to conserve the indigenous bee.[176]

However, I would caution against importing to its former habitat even a black bee that, according to some ideal or model, has been reassembled from the genetics still available in some other locality. This could add up to being almost as unsustainable as importing Italian bees into northern Europe. Indeed, even attempting to reinstate it from surviving local stocks may not offer a durable solution, except in regions whose geography would enable the establishment of a drone population that would help maintain the purity of the desired racial characteristics.

Finally there is the matter of natural selection. If bees are domesticated at all, they are only partially so. They can be more or less exposed to the wise invisible hand of natural selection depending on the degree of intervention chosen by the beekeeper. In the natural swarming process, many queens are raised, but only one goes into requeening the parent colony after the old queen leaves. One or two more might head secondary swarms or casts. But queen breeders would regard it wasteful if, say, out of 10 queen cells with apparently normal looking queens, only one or two were to be retained. Thus, artificial breeding short circuits an important step in natural selection which involves a number of quite complex behaviours between the competing queens, and between the queens and the workers.

The other array of complex behaviours that is short-circuited by breeding with disregard for swarming is the whole process of establishing new nest sites. This involves swarming; clustering; deciding on which new site to go to; navigating to it; occupying it and constructing a nest in it sufficiently provisioned to survive winter. One wonders how many generations of preventing this process from running its full course will so weaken the bees genetically that they will no longer be fit enough to survive without human intervention. In one study it was found that about 80% of swarms do not make it through the first winter.[177] This is a harsh selection process, which, if removed by too much care from the beekeeper, for example routine autumn feeding, simultaneously removes a stimulus to long term fitness in the bee population.

In situ breeding with the whole colony

To maximise the intimacy of new queen and colony by raising queens from normal queen cells, it follows that we must breed queens while working with the swarm impulse. This involves breeding being largely conducted by individual beekeepers in their own apiaries using locally adapted bees.

Artificial swarming

Many beekeepers have apiaries in residential areas and thus, unless they have very understanding neighbours, often prefer not to let their colonies swarm naturally. An option therefore is to carry out some kind of artificial swarming to pre-empt the natural swarming which would otherwise occur.

Let us pause here to consider the overall situation: beekeepers keep bees in urban environments, knowing that a powerful instinct of the bee is to swarm. Beekeepers then expect their bees to conform to the standards of urban environments even though the bees did not ask to be kept there. With the growing public interest in helping the honey bee, and the corresponding increase in urban beekeeping, it appears that some form of public education is called for regarding what is happening when bees are swarming, a situation where bees are generally at their most docile.

To return to artificial swarming: bearing in mind that we are here interested in queens raised in normal queen cells, once occupied queen cells are present, a colony is split, whilst ensuring that the part without the queen has one or more queen cells. To know when a colony is ripe for artificial swarming requires multiple inspections ahead of the manipulation. The increased intrusion this involves may have consequences for colony health.

Conventional practice requires removing all but one or two cells from a colony that is split in the process of artificial swarming. However, this risks selecting a less vigorous queen or, worse still, ending up with no queen at all. Bees have managed to select their own queens for millennia without our help, so should we not let the maturing queen and the bees decide? The answer partly depends on whether the beekeeper veers towards instrumentalisation of bees, i.e. wanting colonies to be under his total control for his own profit and convenience – in which case he would probably have not let the swarming process even start – or whether he sees the bee as also having its own intrinsic value as part of the natural landscape. In the latter case it would be better not to interfere too much and let the bee adjust its genetics to what is coming towards it in terms of changes in available forage, climate, pests and diseases. Too much artificial selection risks reducing the genetic variability and thus the future options for tackling new challenges. However, account has to be taken of the fact that leaving the full complement of queen cells risks swarms emerging despite the split.

It is worth examining what traits are generally selected for in breeding programmes and considering whether they are really worthwhile in the context of relatively natural beekeeping. The following list does not exhaust the possibilities:[178]

Docility (not jumping, stinging, following, biting or head butting).
Steadiness on the comb (not running).
Propolis (not too much).
Cool air clustering on frames ('drippy bees').
Queen balling.
Brood area, cell size and pattern (compactness of brood, no empty cells in brood
    patches).
Pollen supply (pollen packed over, around and under the brood nest).
Comb building (speed in occupying supers, drawing foundation, honey storage, and
    quality of comb capping).
Early spring build up.
Thriftiness.
Hygienic behaviour and house cleaning (corpse and debris removal; polishing).
Chalk/sac brood (not present).
Grooming and mite damaging.
Queen-right supersedure.
Morphometric data for subspecies selection (discoidal shift, cubital index, colour,
    overhairs, tongues, tomenta).

The first five are mainly for the convenience of the beekeeper who has to open colonies and take out combs relatively frequently. Even if he is not being stung he does not want to be distracted by head butting bees or followed back to his house or vehicle. But the behaviours largely relate to the bee protecting its home from intrusion or disease. Most of the remaining behaviours could be subsumed under questions which can be answered without opening the hive, namely is the colony building up as one would expect and does it give a surplus of honey at harvest time in an average melliferous season? So rather than reducing selection to its smallest components, which is no doubt entertaining for the breeder, it could be tackled more holistically.

Clearly, even if we avoid intensively targeted selection for the various queen traits that have appealed to beekeepers from time to time, it does not mean abandoning selection altogether. If, despite good foraging conditions, a colony is not developing normally in comparison with others, then it can be culled or, if it is not diseased, it can be united with another. Indeed, it is a good policy to get rid of weak colonies by choosing which queen to kill or letting them fight it out. One strong colony produces more honey than two weak ones. And if a colony available for breeding shows undesirable behaviours outside the normal range, for example it is so defensive that it stings the neighbours, then it can simply be eliminated from any further breeding and, in extreme cases, re-queened.

However, we have found that colonies kept under minimal stress, especially avoiding the stress of frequent hive opening, are far less inclined to defensive behaviour. Indeed, selecting for docility is a task created by a particular type of beekeeping. And if defensiveness is linked to other, as yet undiscovered traits that might be beneficial to colony health, is it wise to actively select for docility?

Even in skep beekeeping a fair measure of holistic selection – i.e. whole colony selection – was practised. It took place when deciding on which colonies to allow to continue to the following season.[179] The hives were categorised according to comb development, both quantitative and qualitative, as well as hive weight. The lightest had their bees driven into hives where they were needed, and the honey was harvested or the comb saved for the following year. These light colonies would be those that had failed to build up as well as average, indicating deficient queens. The hives of medium weight would become the following year's stocks. These may or may not be supplemented with bees driven from the heaviest hives, called the honey hives, which as their name suggests were those taken for harvest. Harvesting from skeps, unlike from top-bar hives, generally involves dismantling the entire colony. First the bees are driven out to avoid the wasteful and barbaric practice of asphyxiating them with sulphur.

Fig. 8.3: Queen cell. Photo: Tim Walsh,
greenroadfarm.com

Working with the swarming impulse whilst retaining a high degree of control over the outcome is of course more convenient with frame hives because of the relative ease of access for inspection and subsequent manipulation. However, colonies in top-bar hives can be split just as they can in frame hives. Even so, it should be obvious that neither horizontal nor vertical top-bar hives easily lend themselves to intensification of queen breeding, i.e. raising many queens from a single colony. In the horizontal (long) top-bar hive a split for raising a new queen is relatively easy as the hive can be simply divided and rotated to put the queenless part with the foragers and queen cells in one half whilst leaving the other half with the old queen.[180] In a vertical top-bar hive such as a Warré hive, splits generally involve working with whole boxes, because the combs are not so easily moved about as in a horizontal top-bar hive which, from its very nature, undergoes regular comb management.

Gilles Denis, who maintains several hundred colonies in modified Warré hives, keeps up the numbers with such splits, with or without finding the queen.[181] Additionally, for queen rearing, he uses semi-frames (see Chapter 10).[182]

Incidentally, both kinds of top-bar hive, vertical and horizontal, contain no queen excluder. The queen is in principle free to wander over all the comb. Could this freedom facilitate the intimate relationship between colony and queen to which

Steiner was referring?

Natural swarming

The splits described above result in artificial swarms and, even though they use queen cells that are part of the natural swarming process, they nevertheless involve a degree of interference in the normal process of colony reproduction. Like a colony, a natural swarm also has a more intimate relationship to its queen than is likely to be produced in an artificial swarm. Swarm preparations, including recruiting the departing bees, start weeks in advance. At the very least, this swarm-queen relationship comprises the quantity and mix of bees, i.e. the ages of the bees and thus their stages of development and corresponding functions in the colony. I have been struck by the speed with which a natural swarm fills an empty Warré hive box with comb and builds up a vigorous colony. For example, Timothy Malfroy (New South Wales) shook swarmed a colony into a new Warré hive in late summer and within two weeks it was at work on its fourth box of comb.

I prefer to start new top-bar colonies with natural swarms. But this is clearly not an option for everyone. It requires much more vigilance at the apiary and in non-rural areas risks annoying neighbours. Even when a swarm issues, it is not always practicable or safe to take it, for instance if it is high in a tree. So in order to minimise losses of swarms, some inexpensive bait hives can be distributed in the vicinity of the apiary and inspected frequently. To work well, bait hives should be a few metres off the ground in a prominent location; be about 40 litres capacity; have had bees in before or smell of bees; have an entrance of about 12 square centimetres at the bottom; face southward; and be over 300 metres from the apiary.[183,184]

To make a new hive smell of bees, the inside can be coated with beeswax and/or propolis and the entrance area with propolis. An alternative solution is to hang a piece of disease-free comb inside that has been used at least once for brood. The attractiveness can be increased by sprinkling a few drops of lemon grass (*Cymbopogon citratus*) essential oil inside. This has been found to be 2.5 times more effective than beeswax alone.[185] The same researchers found that synthetic Nasonov pheromone was about half as attractive again compared with lemon grass oil. Some beekeepers recommend ensuring that the area round the entrance is white so as to maximise the contrast for attracting scouts. This can be done by painting or pinning on white card.

Provided bait hives are checked regularly, there is no need to use actual hives for this purpose, a cardboard box protected from rain will suffice. Indeed, bait hives have been made from all sorts of containers. An ingenious solution involves a piece of drainage pipe from which the swarm can be simply lifted and transferred to a

Warré hive.

Fig. 8.4: Drainpipe bait hive. Photo: Steve Ham, Spain

I accept that allowing breeding through natural swarming places additional demands on the beekeeper, although it is nowhere near as demanding as lambing time is for the sheep farmers in my locality who are often awake half the night to attend to their ewes. Yet I concur with Erik Berrevoets who suggests that 'perhaps beekeepers who are unable to tend their colonies during this period of the year should reconsider their involvement with beekeeping'.[186]

Some swarms will nevertheless be lost. But in what sense is this a loss? It certainly reduces the profitability of beekeeping, but it also has the potential to restock the landscape with feral colonies, most of which, at least in the UK, were wiped out by Varroa. Thus an apparent loss is partly mitigated by the contribution to the surroundings, even possibly to other beekeepers, thus helping the maintenance of genetic diversity.

9. THE PEOPLE'S HIVE OF ABBÉ ÉMILE WARRÉ

Each winter, all my childhood friends ate an abundance of delicious bread and honey, just as I did. Twenty years later, I was the only person who had beehives. In some gardens, there was an abandoned Dadant or Layens hive, empty of course. The owners had let themselves be tempted by the advertisement of some on displays at agricultural shows. They believed they would do better with these modern hives. In fact they abandoned the only hive that suited them. [...] At my parent's home there was always plenty of honey for masters and workers, even for the farmyard animals. All our friends in the village also had their share each year. Warré (2010)[187]

And to corroborate Warré's comments above about the decline in beekeeping, here is a quotation from a UK government report in 2008:

100 years ago there were around 1 million bee hives; this had reduced to 400,000 in the 1950s and further reduced to the 274,000 today.[188]

Much of the foregoing has been written with top-bar hives in mind because they are in many ways more manageable and easier to construct than skeps, and are arguably more bee-appropriate than frame hives. Of the top-bar hives, I prefer the vertical top-bar hive for its lower requirement for intervention by the beekeeper whilst functioning optimally.

The essential principle of the vertical top-bar hive is that new boxes are always nadired, i.e. placed under the brood nest, and honey is always harvested from the top. The brood nest continues to grow downwards as in the cavity of a feral colony.[189] Vacated brood cells are filled with honey at the top. As there are no supers, a major stimulus to the bees to overproduce honey is eliminated and with it is removed a potentially stressing factor. The bees have a strong urge to fill any space above the brood nest with honey. This urge is especially exploited in the practice of 'chequerboarding' which creates a mix of empty and filled combs above the brood nest so as to break up the honey band with the ultimate aim of distracting the bees from swarming.[190] But in a Warré hive, without any supers at all, only so much honey can be stored at the top as cells are vacated by brood.

This necessarily limits the honey crop, which could average as little as half that of a supered colony in the same locality. It is worth noting that in a dissection of eight thriving feral nests taken in July/Aug in New York state, USA, an average of 15 kg honey was found to be present although maxima as high as 200 kg have been reported.[191]

The first vertical top-bar hive to be brought to my attention was that of Johann Ludwig Christ (1739-1813),[192] followed shortly after by that of Abbé Émile Warré (1867-1951).[193] The final edition of Warré's book *L'Apiculture pour Tous*, the edition that put the greatest emphasis on the top-bar version of his 'People's Hive', was available only in French at the time, and that availability thanks to the work of Guillaume Fontaine in scanning an out of print copy for publication on the Internet.[194]

Fig. 9.1: Christ's hive as reconstructed by Rudi Maurer. Photo: Rudi Maurer

Warré's book seemed to me worth translating into English, not only for details of the hive's construction and management but also for Warré's many observations relevant to our theme here. Furthermore, the hive appeared to be so easy to make and run that it could easily result in beekeeping becoming as commonplace as it used to be in the 19th century. Furthermore, the hive already had a significant following in France where there were also in use several contemporary modifications which we discuss in Chapter 10. My contacts in France soon proved to be an enormous help in introducing the hive to anglophone beekeepers. However, I am still puzzled that this did not happen half a century ago. If any reader discovers that the hive found its way to English-speaking countries earlier than 2006, I should be most interested to hear about it.

Publication of *Beekeeping for All* on the Internet was accompanied by the appearance of articles on the hive in beekeeping journals in the United Kingdom, USA, Canada, Australia, New Zealand and Mexico and followed by the formation in 2007 of an e-group.[195] There were already several such groups in France together with a number of French websites for the hive and its modifications.[196] The e-groups are excellent places to get help with questions regarding hive construction and management, as well as reliable sources of bees in your region.

Precursors

With the growing interest in the hive amongst anglophone beekeepers in 2008 and 2009, several further precursors of Warré's concept were brought to my attention. Indeed, Warré acknowledged that the People's Hive is not a revolution in beekeeping'.[197] After he had developed it, beekeepers told him about the hives of Palteau and DuCouédic based on similar principles. Other precursors that are now known to me are:[198]

> John Gedde (1647-1697), book published 1677 (3rd edition).
> Guillaume Louis Formanoir de Palteau (1712-?), book published in 1756.
> Samuel Linnaeus (1718-1797), book published in 1768.
> Thomas Wildman (1734-1781), book published 1768.
> Pfarrer Johann Ludwig Christ (1739-1813), book published in 1779.
> Bryan l'Anson Bromwich (? - 1805), book published in 1783.
> Pierre Louis DuCouédic de Villeneuve (1743-1822), book published in 1813.
> Edward Bevan (1770-1860), book published in 1827.
> T. M. Howatson (?-?), book published in 1827.
> Nicolai Vitvitsky (1764-1853), book first published in 1846.

Illarion Semenovich Kullanda (1848-1922); book published in 1882; a near precursor of Warré.

Not all the above are top-bar hives but they all follow the basic concept of nadiring new boxes and harvesting at the top. For example, the hive of Linnaeus is a skep with nadirs. This was common practice in skep beekeeping at least until the end of the 19th century.[199] We will not review the various versions here. The Warré hive itself, together with its modifications, including those that depart from the square footprint, suffices for most purposes for illustrating the principles and practice of this beekeeping concept.

Construction

Fig. 9.2: Exploded view of the People's Hive of Abbé Émile Warré. Note that there is an additional cloth, not visible in the photo, that serves to retain the quilt contents.

Although the People's Hive is easy to construct, for various reasons some beekeepers may wish to buy them ready made either in kit form or assembled. I try to keep an updated list of suppliers on the English Warré beekeeping web portal.[200]

This chapter is not intended to be a substitute for reading *Beekeeping for All*. On the contrary, I thoroughly recommend studying it before embarking on Warré beekeeping or trying any modifications of the People's Hive. Warré's book is aimed at beginners, but there are many aspects of the evolution and running of the People's Hive which will interest more experienced beekeepers. I also strongly recommend that beginners join a local beekeeping association where a great wealth of experience can be found.[201]

A Warré hive is a tiered top-bar fixed-comb hive comprising a stack of at least two boxes. Boxes are butt jointed and nailed at the corners, each of internal dimensions 300 x 300 x 210 (deep) mm with eight 9 x 24 mm top-bars at 36 mm centres. The floor, a plain board, is notched to form a 120 mm wide entrance and has an alighting board nailed underneath. The depth of the notch allows the bees to pass under the front rim of the bottom box.

Fig. 9.3: Floor

The internal dimensions of the box resulted from long researches involving the construction of some 350 hives, but are essentially developed from features, such as cavity size and shape as well as the number and dimensions of combs, embodied in the hives of Abbé Voirnot and Georges de Layens.

Fig. 9.4: Warré box top

The box walls are at least 20 mm thick. Warré preferred 24 mm. I use 25 mm, but finding wood this precise thickness is not easy. The top-bars rest in 10 x 10 mm rebates, but, to ease construction, can just as securely rest on battens nailed 10 mm below the box rim, although this hampers comb removal for inspection if it is ever required. The bars have a bead of wax or starter-strip fixed in the middle

Fig. 9.5: Top-bars underside showing wax starter strips

of their rough-sawn undersides.[202] Frèrès and Guillaume recommend applying linseed oil or petroleum jelly on the planed upper surfaces to minimise adhesions from combs in the next box up. They also put a 2 x 2 mm groove in the centre of the undersides of the top-bars, but this is not necessary.[203] The bars are secured with very thin headless pins, such as 20 x 1 mm Japanned frame pins with the heads cut off. The pins should not project too far into the rebate, otherwise bar removal is difficult. A depth of 5 mm is sufficient. Alternatives to pins are discussed in the next chapter. Each box has ample, firm handles capable of supporting a fully stocked hive.

On the top box rests a layer of coarse-weave hessian sacking (burlap) stiffened with rye flour paste to prevent the bees fraying it.[204] The sacking can be cut from used coffee sacks or from peanut sacks obtained from a pet food shop. Above that is a 100 mm deep box, the *coussin* which is translated as 'quilt', as this term conveys its function better and is not unfamiliar in hive construction. The underside of the quilt is covered with plain sacking and the top left open so it can be inspected and the filling renewed as necessary. It is filled with natural insulating material such as wood shavings, sawdust, straw or dried leaves etc. Apart from its insulating function, this helps to control humidity through absorbing excess moisture onto the large area of hydrophilic surface and thence into the cellulose matrix. As we have discussed, this probably has a humidity buffering function.

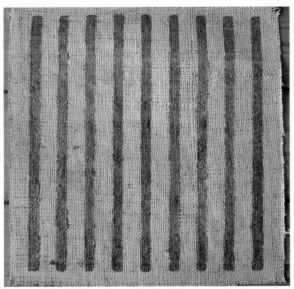

Fig. 9.6: Top-bar cloth that has been in use, showing propolis

Fig. 9.7: Quilt filled with wood shavings retained by sacking

Resting on the quilt is a wooden ridged roof containing a board to keep mice out of the quilt, and a ventilated cavity, which reduces solar heating of the top of the hive.

Fig. 9.8: Warré roof underside

Appendix 2 contains working drawings for constructing the hive. Plans for printing out in both metric and imperial measurements are available as PDF files.[205] Please bear in mind that Warré designed the People's Hive to be constructable by those with rudimentary woodworking skills and basic equipment. Anyone who is completely new to woodworking may like to read Nick Hampshire's Warré hive construction guide.[206] However, this contains some departures from Warré's design.

Warré recommended painting the hive with white oil-based paint. Many beekeepers now prefer to use latex-based paints because of their weathering and 'breathing' properties. I use oil-based paint on the roof, but give the boxes two coats of heated raw linseed oil with at least twenty-four hours drying in between. Pure tung oil has also been used with success, though it requires a long drying period.

Management

Apiary site

Common sites include gardens, city rooftops, allotments, field margins (livestock fenced), wasteland etc. Walls, fences, hedges and/or screening nets (windbreak) help funnel flight traffic in the desired direction. The flight path near the hive entrance should not point immediately across thoroughfares or any place where people frequently pass. The entrance should ideally face anywhere between east and south to rouse the colony to foraging at sunrise. Unless the supply of forage is better than average, try to limit each site to about three hives to avoid stressing the bees by over competition for resources. If there is another apiary of significant size close by, then consider reducing the number of hives to less than three.

At all my hive positions, with the help of a spirit-level I set in the ground a recycled paving slab so that it is not only horizontal but also its upper surface is level with the surrounding ground. This gives a very stable base on which to place the hive stand. I generally face the hives in the quadrant between east and south, so this needs to be considered when setting the paving slab. Some further comments on legs and stands are included in Chapter 10.

Use of smoke

I use smoke very sparingly, just enough to complete the desired manipulation and in many cases not at all. Given that smoke contains a cocktail of toxic chemicals and is obviously irritating to the bees, it is advisable not to use too much of it. Another disadvantage of smoke is that it disrupts colony communication and it may take some time fully to restore the pheromonal structure of the colony. Even so, smoke from natural products is probably preferable to resorting to some of the synthetic smoke substitute sprays offered to beekeepers.

Warré advised smoking the entrance before every manipulation, but I find that this is generally unnecessary. For example, if a box is nadired with the help of a lift, the bees seem unconcerned and the new box can be slid across the floor, carefully displacing any bees standing there. In some situations, such as hiving a swarm or a package, a spray of weak sugar solution is helpful.

Fig. 9.9: Five-box Warré hive on stand on concrete slab

Fig. 9.10: Nadiring a box with a Warré hive lift

Hiving bees

Warré advised that a natural prime swarm of 2 kg gives best results. However, a 1 kg bought package of bees – an artificial swarm with a queen – has been found to work very satisfactorily by Warré beekeepers in North America. If you would rather not pay for bees, then tell your local beekeeping association, police, pest control department and fire station that you will take swarms. It is also well worth setting up bait hives (see Chapter 8).

Fig. 9.11: Commercial package of bees. Photo: Larry Garrett

Ideally try to hive bees just before or during a main nectar flow. This gives the best conditions for comb and colony development and saves having to feed. In the UK this would usually be some time in May. Late April is possible in a good spring. By the end of June it is starting to get too late for good colony build up without copious feeding.

A natural swarm can be shaken or brushed into an inverted top box of a 2-box Warré and the inverted box then set the right way up, or the bees are run up a board into a Warré hive that is already set up. The workers from a package can be run in too, in which case release the queen from her cage at the hive entrance once the inrush begins.

A commercial package can be hived by direct release of the queen as follows: have at the ready a 2-box Warré and a third box without bars to use as a funnel. Open the queen cage and place it on the hive floor. Immediately, put on the floor the two boxes with top-bars and the third (funnel box) on top. Shake the bees from the package into the Warré. Lightly smoke the bees from the top box down into the hive if necessary and replace it with the top-bar cloth, quilt and roof. If any bees are still in the package, put it beside and touching the alighting board.

Fig. 9.12: Running a swarm into a Warré hive

An alternative method is to place on the hive floor a couple of small sticks about a bee space thick then put one box without top-bars on the floor. Remove the feed can and queen cage from the package and pour most of the bees into the hive box. Place the package with its remaining bees on end on the sticks, taking care not to crush bees. Place two boxes with top-bars on the bottom box. Remove the plug over the candy in the exit hole of the queen cage and attach it to the bars of the top box so that the mesh is not covered. Close up the hive. On the following day, when the cluster has formed round the queen above, remove the bottom box, package and sticks. If pollen is not coming in after a couple of days, it is as well to check that the bees have released the queen.

Enough bees and the queen may be transferred from a frame hive or other type of hive by driving, i.e. drumming with sticks in the traditional way.[207] This has the advantage of simulating swarming in that the brood, with any pests (mites) and diseases it may contain, is left behind and, if required, can be allowed to requeen itself. And when the parent colony is not to be continued, shook swarming from frames of combs directly into a Warré using a funnel has the same result. These methods of transferring adult bees to a Warré hive are less likely to be successful with a nucleus because of its smaller population of bees, although given warm weather and favourable foraging conditions, they are not out of the question.

Fig. 9.13: Funnel for transferring bees from frames to a Warré hive

One can transfer the combs and bees of a frame hive colony, or a feral/wild colony in a building or other site where it is not wanted. This can be done in one of three ways: either cutting out combs and tying them to top bars, supporting the combs in split frames that open like a book and fit a Warré box, or propping the combs upright with sticks as spacers in a Warré box to the bottom of which is stapled chicken wire netting.[208]

Several Warré beekeepers have successfully transferred frame hive colonies to Warré hives by placing them in a suitable box on an adapter board on top of the Warré hive. This would certainly be an option to consider if one only has access to bees as nucs, although it is a very slow method and does not always succeed in the first season. It helps if the bottom bars are removed from the frames and they are supported in such a way that the comb is only a bee space from the top-bars of the Warré hive.

Fig. 9.14: Langstroth 8-frame box on a Warré ready for a Langstroth nuc transfer.
Photo: Larry Garrett

For a bottom bee space hive, such as a National, the top of the adapter board should be flush with the top of the Warré box. It can be supported by the Warré box handles. Those who are confident in finding the queen may wish to speed the transfer of the colony to the Warré hive by placing the queen in the top box of the Warré hive below a queen excluder which rests on the adapter board. Ideally, she should be placed on a piece of worker comb securely fixed to one of the central top bars so that she can resume laying immediately. Queen transfer may be safely carried out with a clip-type queen catcher. As brood hatches out of the combs in the nucleus or frame hive, they are filled with honey and the top box can be treated as a super.

Fig. 9.15: Langstroth hive transition to a modified Warré hive.
Note the provision for entrances in each box. Photo: Bill Wood

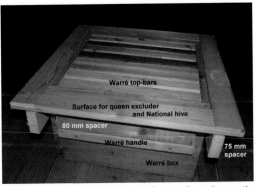

Fig. 9.16: Frame hive to Warré hive adapter that sits on the Warré box
handles

Feeding

If there is no nectar flow after hiving, feed with your own honey syrup (2:1 honey:water by weight), or if no honey is available, use syrup comprising 1 kg refined granulated sugar dissolved in 500 ml water. If the weather is consistently warm, put it in an open container loosely filled with straw or with a layer of wine corks (to stop the bees drowning in it) on the hive floor with an empty box round it. If the weather is cold, it is better to use a top feeder, preferably a contact feeder which can be as small as a honey jar. Warré describes the construction of two types of feeder, a floor feeder for spring and top feeder for autumn.[209]

Fig. 9.17 Floor feeder for spring. Photo: Cernagor Nicolae, Romania

The larger, autumn feeder is based on a hive body box. Bees enter it from below and access the surface of the feed via a gallery covered with glass or Perspex (acrylic sheet). When the liquid level has dropped to the bottom of the tank, the bees can enter the tank and lick it dry. Both feeders can be used for feeding comb honey.

Bag feeders have also been tried successfully during installing colonies. If the feeder is installed at the top, say in a shallow eke, it is important to have no space above the bag big enough to allow the colony to cluster and start comb there. In fact there have been a number of disastrous results with bag feeders in Warré hives. Unless the user is very confident about their use in a Warré hive, they are best avoided.

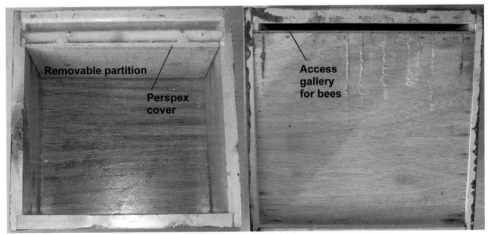

Fig. 9.18 Top feeder for autumn. Left: Top view; Right; Bottom view

Monitoring progress

Storch details how a lot can be learnt from entrance activity.[210] All is well if, on rainless days, the bees are coming and going, many carrying pollen. The first box can fill with comb in a fortnight under UK conditions, sometimes less. In about another fortnight, though always ahead of demand, you can add a third box underneath. To see how far comb building has progressed, slide the hive backwards a little on the floor to make a satisfactory opening and look upwards with a torch, or use a mirror. Do not do this often. Part of the point of Warré beekeeping is to leave the bees alone.

Fig. 9.19: A Warré hive colony just starting its second box; viewed from below

There will come a time when a 2-box hive is too heavy for one person to lift, for example when inserting a new box. If an assistant is not available, then rather than dismantling the hive box by box and putting them on temporary stands, which goes against the whole point of the hive, a lift can be used. Hive lifts are commercially available, but it is fairly easy to construct a fork-lift out of scrap materials that picks the whole hive up by the handles once the roof is removed. A windlass is mounted on a frame and connected with pulleys to a board which slides up and down in grooves. The board has the two tines of the fork firmly mounted on it. These are slid under the bottom box handles and the hive lifted. The action is very smooth and generally seems to go unnoticed by the bees. The construction of an example fork lift is shown in Appendix 3.

Harvesting honey

No bee escapes, blowers, fume boards, chemical repellents, extra supers, uncapping knives, extractors or bottling tanks are needed when harvesting honey from a top-bar hive. This hugely reduces the ecological footprint of beekeeping. The essential equipment, knives, strainers and bowls, are found in an ordinary kitchen. With the horizontal format top-bar hive, bees are simply brushed off the combs and the comb cut into a bucket with a lid. Honey is drained or squeezed from crushed comb. This avoids the exposure of fine droplets of honey to air as occurs in extractors and helps conserve the subtler aspects of flavour and bouquet.

Under UK conditions, if you have populated your Warré hive in mid-spring and had an average summer, you might be able to harvest a box of honey in early autumn when the nectar flow largely ends. Remove the roof and quilt. Peel back the top-bar cloth and smoke any bees down into the second box. Loosen the box with a hive tool, if necessary with a gentle twist of a few degrees each way in the horizontal plane. Check that there is no brood by tilting the box and looking underneath into the combs. They can be gently parted with the fingers if necessary. If brood is seen, replace the box. Check that the next box has at least the equivalent of six combs of honey (about 12 kg.). This is the winter stores. If, based on knowledge of local conditions, there is a further nectar flow that can generally be relied on, for example from 'Impatiens glandulifera (Himalayan balsam), Reynoutria japonica (Japanese knotweed) and Hedera helix (ivy), then a little less than six combs of stores would be sufficient to leave at this stage.

If there are still quite a few bees in the harvested box, they can be driven out by a combination of smoking and drumming with sticks. If a cluster of bees is very reluctant to leave, then suspect that the queen is with them and temporarily return the box to the hive with great care, unless you are able to catch her safely and return her individually.

If there are open honey cells at the bottom of the box, precautions must be taken to avoid spilling honey, which could trigger robbing. One solution is to temporarily replace the box on the hive with a gap of a few millimetres between it and the hive using wedges. It is left there for a short while until the bees have cleaned the honey out of the broken cells. If the box is raised more than 3 mm there is a risk of robbing, but this can be prevented by tying a piece of cloth round the gap. Bees may have re-entered the box after this procedure so will need to be driven out again.

A common way of clearing bees from supers is to put a bee escape between the boxes to be removed and the colony below. However, in a Warré hive, where the queen is in principle free to roam the whole cavity, it is important to be sure that she is below the escape board, i.e. down in the brood nest.

Another way to remove the bees, which is safe to do only in a good nectar flow, when robbing is unlikely, is to leave the box a little distance from the apiary for the bees to fly home.

Before closing up the hive finally, renew the top-bar cloth and, if it is mouldy, the quilt filling.

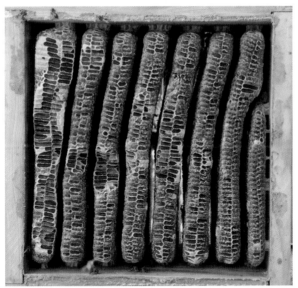

Fig. 9.20: Warré box full of honey. Photo: Steve Ham, Spain

Extracting honey

The lightest coloured combs at the top corners of the box are suitable for harvesting as cut honeycomb because they will not have had brood in. The rest is sorted to remove any remaining significant areas of pollen, cut up small or shredded, and drained through a sieve. The drained wax can either be pressed dry in a cloth, or washed with lukewarm water and the honey syrup either fed back to the bees or made into mead.

A number of different types of press are used by Warré beekeepers. Most presses are capable of handling whole combs.[211]

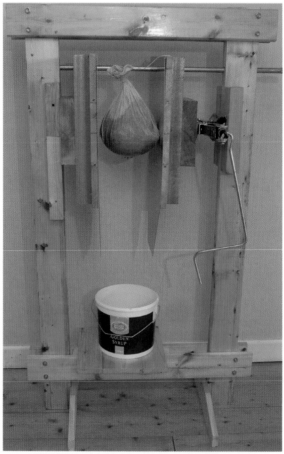

Fig. 9.21: A home-made press based on a
car-jack and two sheets of stainless steel[212]

The objection has been raised that honey harvested from the Warré hive comes from combs that have had brood in them. The argument goes that the cells are lined with cocoons and faeces of young bees and this is therefore unhygienic. Against this objection can be set a number of observations. Firstly, the cells of honeycombs are cleaned out scrupulously by the bees before being used again. Secondly, honey from brood comb has been consumed by humans without ill effect for hundreds of thousands of years; it still is in many parts of the world where skep and other more natural forms of beekeeping are practised. Thirdly, even in frame hives, which are managed to harvest honey only from comb in supers where there has never been brood, the incoming nectar usually spends part of its time in brood comb at the bottom of the hive. Furthermore, I note that one beekeeper of sixty years experience used old framed brood comb to go to the heather.[213] Thus his comb performed a final service before it went to the press to extract the thixotropic honey from it.

Rendering wax

It helps to wash the wax until it is relatively free of honey, even if it has been pressed to a cake. It is steeped in warm water and rinsed a few times before leaving to dry for several days on a cloth supported by a mesh tray, or open weave basket. Wax that has not been pressed will yield much honey that can be fed back to the bees in a top or floor feeder.

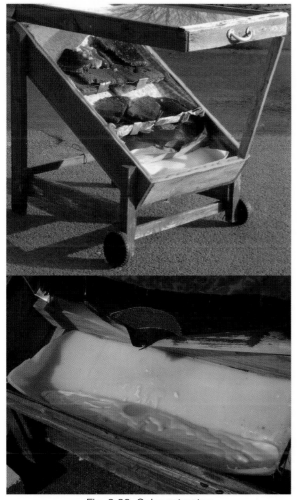

Fig. 9.22: Solar extractor

Wax is rendered in the most sustainable way in a solar extractor.[214] If the bottom of the extractor is lined with an old cotton sheet to act as a filter, wax is collected of high enough quality not to need further processing. Any drips of caramelised honey usually settle to the bottom of the wax cake and can be rinsed out in warm water or, if embedded in the wax, cut out with a knife. Apart from selling wax directly, value can be added to it in a number of ways, for example by making it into candles. This is done by melting it in a tall can in a *bain marie* and dipping candlewick obtainable from any candle-makers' supplies until the desired thickness is reached.

Swarming

As we have discussed, suppressing swarming, which is part of the natural reproduction of the honey bee, risks harming the long term fitness of the bee population. We can let colonies swarm and catch them, if necessary by using bait hives (Chapter 8), and use them for starting new colonies. But this is not an option if there is any chance of annoying neighbours. In that case, they may be split before swarming time in the second year, or artificially swarmed as Warré describes.[215]

Fig. 9.23a: Swarm on the move

Fig. 9.23b: Swarm arrived at bait hive

Fig. 9.23c: Swarm entering bait hive, bees fanning and scenting

Fig. 9.23d: Twenty-five minutes later, most bees have entered the bait hive

Fig. 9.23e: Half an hour later, a few stragglers left at the entrance

Varroa control

Beekeepers generally use chemicals against Varroa mites, but this is not sustainable in the long term and adversely affects bee health. Honeybees and mites will eventually co-adapt, co-evolve. Any intervention postpones this. I do not treat my Warré colonies as they create ideal conditions that help bees control mites themselves. At the time of writing, the survival rate of my Warré colonies in their fourth season is 33%, with the losses probably being attributable to three seasons in succession of poor forage (2007-9). But I nevertheless risk continuing to lose some colonies to Varroa, even in good seasons. If you don't want to take this risk, for example because you have only one or two hives, consider dusting the bees with icing sugar (powdered sugar): intrusive and messy, but it is said to give some control provided that it is done frequently and thoroughly. A simple dusting dispenser is described by Dennis Murrell.[216] However, see also Chapter 7 (Varroa) for a comment on dusting.

Fig. 9.24: Pollen loads: encouraging signs in early April at the entrance of a Warré hive in its fourth season without Varroa treatment

Wintering

If there has been a normal season and the advice to leave 12 kg stores has been followed, then the final tasks of the season are to ensure that the bees have two boxes of comb to winter on, and to place a suitable mouse guard on the entrance. Mouse guards can be improvised or bought. The main requirement is that the guard leaves a slit no higher than 7.5 mm or comprises a series of holes no greater than 9 mm diameter. Warré describes a mouse guard which ingeniously doubles as a robber guard and can be quickly fitted and switched between the two functions.[217]

If the season has been abnormally poor, it may happen that there are less than

12 kg stores available to the bees. In which case it may be necessary to feed (see 'Feeding' above). This has to be done while it is still warm enough for the bees to access the feed and store it properly, which in the UK is generally no later than early October.

An alternative way of assessing the weight of stores is to weigh the hive. The simplest way to do this is with a spring balance hooked under each handle in turn of the bottom box of the pair to be weighed, with the quilt, but not the top-bar cloth removed. The weights are added and from the sum is subtracted the weights of two boxes with top bars together with an allowance for the weight of the empty comb and bees. I use a 30 kg spring balance which is supported and lifted by means of the screw on a joiner's sash cramp. However, a wooden post with a pivoted lever on it works just as well. The box must be separated from any propolis that is sealing it to the floor and need only be lifted a few millimetres until it is obviously suspended.

Fig. 9.25: Hefting a Warré hive with a spring balance. A: hive, sash cramp, balance, chain and hooks; B: detail of lifting screw, top hook and balance; C: detail of chain and hook under box handle

Fig. 9.26: Putting candy in an eke on a Warré hive. A: Shallow eke lined with plastic bag used as mould for candy. B: Top-bar cloth peeled back as Perspex sheet is slid on in its place. C: Eke of candy is inverted on a thin sheet of metal, placed on the hive and the metal and Perspex slid away

If, from hefting in the late winter, there is evidence that stores are getting low, then the option is available of feeding candy or fondant. Recipes for this abound. The general principle is to completely dissolve refined sugar in the least amount of water and boil until the temperature reaches 117 °C or 'soft ball' on a cooking thermometer. Allow to cool while stirring and when crystallisation starts, immediately pour into moulds.

For a large amount, say 2 kg, it can be poured into an eke of about 40 mm deep placed on a board. It helps to line the mould first with a polythene sheet. When the candy has set, the eke containing it is inverted on a thin sheet of metal, such as a baking sheet. The whole slab of candy can then be placed on the hive without letting the heat out in the following manner: while rolling back the top-bar cloth at one edge to a maximum of 75 mm, follow the retreating cloth with a sheet of acrylic or glass so that the top bars remain covered throughout. Place the inverted eke with its candy and metal sheet on the top of the hive and gently slide out the metal and glass or acrylic sheets. Replace the quilt and roof. The eke/hive junction will now be exposed to the weather, so wrap a piece of duct tape round it to keep the wind and rain out.

A swarm in the UK may not give a harvest in the first year it is hived. Leaving the stores untouched ensures adequate winter food for the colony, as well as peace of mind for the beekeeper.

Furthermore, this ensures that honey, and not candy or fondant, is available to nourish the new spring brood.

Later years

At a suitable time in the spring before a good nectar flow, which in the UK would normally be in early April, check that the colony survived the winter – a steady flow of pollen entering is a very good sign – remove the mouse guard, clean the floor and add one or two boxes with freshly waxed top-bars. Monitor build-up and ensure that boxes are given ahead of the bees' need for space.

In the second year and beyond, depending on foraging conditions in your area and the season, you might be able to harvest two boxes of honey. But remember that Warré beekeeping means not over-exploiting bees, so they should always be left an adequate amount of their own honey.

10. WARRÉ HIVE MODIFICATIONS AND MODERN MANAGEMENT TIPS

Even while Warré was still alive, other beekeepers had introduced a number of modifications to the People's Hive. Warré took a dim view of them: 'Even the People's Hive has already been a victim of the inventors. They say they are improving it. But the improvements I have heard of are useless, some are harmful, and a few of them absurd'.[218] His judgement of the modifications will no doubt have included the important criterion that he had in mind when designing his hive, namely that it should be simple to construct and use. Most of the modifications I discuss in this chapter admittedly increase the complexity of construction of the hive. However, many simplify its management, and some are certainly essential in climates that differ greatly from those of the Somme and Tours areas of northern France, where Warré developed his hive. And there are a number of other tips included here which, in my view, would be a help to some Warré beekeepers.

Floor and entrance

I use Warré's design for the floor. However, I have made some mesh floors with drawers underneath for monitoring 72-hour Varroa mite drop, as a crude means of checking on the colony's mite burden. These floors are also useful for the ventilation that they provide when moving hives.

Fig. 10.1: Mesh floor

There has been a great deal of discussion amongst members of the Warré beekeeping e-group regarding modifications to the floor.[219] I have not tried any of these modifications, but I present them here for consideration, particularly of the various principles involved. The modifications, several of which may be in a single design are:

1. Mesh for allowing live mites that drop off bees to fall out of the part of the hive occupied by the bees and drop to the ground, or onto a drawer which can, if required, be coated with a sticky substance such as petroleum jelly.

2. Replacing the floor and stand with a box base which forms what several of its proponents call a 'sump'. This may a) have a mesh over it at the level of the bottom of the entrance cut in it; b) be accessible from the back for cleaning and/or c) contain moisture absorbing material such as sawdust.

3. Replacing the floor, and sometimes the stand too, with a box base which could be used for inserting a feeder or a mirror suitably angled in order to view the developing comb in the bottom box, in which case a lamp may also be inserted. It can also be used for inserting a camera. This modification may have a solid floor which can be slid out at the back for feeding and viewing. It also allows cleaning the floor without lifting the hive.

Fig. 10.2: Rear view of sumps. Photo: Steve Ham, Spain

Mesh floors are reported to reduce the number of Varroa in the colony,[220] although some studies have contradicted this (see Chapter 3, section 4). But in combination with a closed drawer underneath they can be used to monitor what is going on in the colony above by studying the various items that drop through the mesh, such as, not only Varroa, but also cappings, pollen loads, old pollen, wax scales and pupal fragments.

There is some justification in the sump concept if hive design is taking its cues from natural nests in hollow trees, which generally have about 20% of the nest height below the entrance.[221] However, I would caution against making any part of a hive that communicates with the brood nest inaccessible to the bees, because it could harbour pests such as wax moths which may breed there and then go on to pose a bigger challenge to the bees above. For example, wax moth cocoons can often be found in the crevices or angles underneath mesh floors. A sump may also make it more difficult for the cleaner and undertaker bees to remove corpses and other potential sources of disease. Furthermore, a sump containing moisture absorbing material could soon become damp and mouldy. However, it is noted that the floors of cavities of feral colonies in hollow trees are completely coated in propolis.[222] But where the bees can gain access to the loose filling in a sump they have been observed to spend a lot of time carrying it out of the hive. Including a mesh screen would prevent this.

Some versions of the sump would be inviting for ants to establish colonies and live off the rich booty to be had above them.

Fig. 10.3: Roger Delon floor

One modification of Warré's own floor that I think is worth considering is to arrange for it to slope down towards the entrance from the back to allow water to drain out. I have had pools of water form on some of my Warré floors. It has never been clear if this is rainwater seepage under the rim of the bottom box or condensation running down the inside walls. I have generally overcome the ponding by tilting the hive *very* slightly forwards with the help of a spirit level. I can tilt in this direction because my combs are aligned cold-way all the year round and thus avoid the risk of comb faces not growing parallel to the sides. However, Roger Delon, who managed 600 modified Warré hives in the Vosges and Jura mountains, arranged for all his floors to slope towards the entrance. Furthermore, he increased the aperture in the summer simply by sliding the hive forward a little on the floor.

In latitudes where there is deep snow for many weeks, usually accompanied by sub-zero temperatures, it is common practice with hives in general to provide small upper entrances for the bees. This is to have an alternative if the lower entrance becomes blocked, which could happen in two ways. One is that there is a risk that the entrance could become blocked with corpses as the undertaker bees are unable to leave the cluster in cold weather. The other is that snow falls could easily block a lower entrance. Upper entrances on Warré hives, formed by drilling a 20 mm hole in the side of the top box, have functioned satisfactorily.

Boxes

An immediately obvious modification is to make the boxes of thicker wood, whilst keeping the internal measurements constant, and size the other parts accordingly. Thicker wood is not necessarily more expensive than, say, 20 mm planed board and provides better insulation against fluctuating temperatures. The thickest tried so far is 50 mm and that was in Alberta. However, to mimic average thicknesses of walls of hollow tree cavities it would have to be 150 mm.[223]

The most widely used modification to the boxes, one that is also available on Warré hives supplied commercially, is to incorporate a window in each box. It is usually placed at the rear. Many of the hives invented by Warré's precursors, going back to John Gedde in the 17th century, have windows in their boxes. The commonest window design for the People's Hive is that of Frères and Guillaume.[224] The window is 4 mm glass sitting in 4 x 7 mm rebates cut in top and bottom window-frame battens at the back of the box. The plans for such a box are given in Appendix 2.

Fig. 10.4: View inside empty Warré box through window

Windows are certainly attractive for beginners as they allow colony development to be monitored relatively easily. I made my first half dozen hives with windows. But if one has an assistant or a hive lift, adding new boxes can be carried out in a timely fashion without the need for windows. An alternative is to put a small viewing hole, say 40 mm diameter, in each box. It is plugged most of the time but comb growth can be checked now and then with a small hand lamp. Some Warré beekeepers even go as far as using these holes as entrances, closing the upper ones as the hive expands.[225]

Fig. 10.5: Warré hive with entrance in each box fitted with disc entrance closers/vents. Photo: David Croteau

The most radical departure from Warré's box design is to increase the number of sides of the boxes. Thus there have appeared a number of modifications that retain Warré's fundamental hive concept – narrow format and nadiring – in which boxes with six, eight, twelve, sixteen and an infinite number of sides have been used, the latter of course being round. This change affects the shape of the floor and quilt but not necessarily the roof.

The internal area has either been kept at 900 square centimetres or adjusted to suit other requirements of the designer. The simplest modification achieves an octagonal format by placing blocks of wood in all four corners of the standard 300 x 300 mm Warré box. This reduces the internal area. But when considered in relation to the fact that cavities of feral colonies in hollow trees have been found to average only 225 mm diameter,[226] the reduction is probably unproblematic. At most it means the brood nest may be spread further vertically.

With boxes of more than four sides, the main construction issue is how to make sound joints at the corners. One solution is interlocking and overlapping corners built up in layers (see illustration of a hexagonal hive). This method is used with a hive design in Hungary that was inspired by John Gedde's octagonal hive.[227] It is managed like a Warré hive with nadiring, although can also be supered.

Fig. 10.6: Hexagonal hive. Photo: Csuja László, Hungary

A more common solution is to use butt joints strengthened internally in some way, either by tongue and groove, false tongues or half-cylinder and socket joints. These all require more advanced woodworking skills and machinery.[228] The box is held together with one or two tight bands while the glue is setting. The rebate for the top-bars is cut with a router.

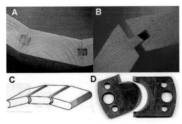

Fig. 10.7: Jointing for rounded Warré boxes: A: false tongue;
B: tongue and groove; C: cylinder and socket; D: cutter for C

With the emergence of Gilbert Veuille's cylindrical hive that is divisible like the People's Hive, hive construction has come almost full circle to that of the skep, in that the composition of the wall is a mixture of straw and plaster. As we discussed in Chapter 3, a cylindrical hive shape is particularly desirable from the point of view of the bees. And Warré writes:

> And the square is the shape that best approaches that of a cylinder, an ideal shape because it favours the distribution of the heat in the inside of the hive. But the cylinder is a shape that is hardly practicable.[229]

The cylinders (*ruchetons*) of Veuille's hive are rigid, and will stand several being stacked vertically, something that is more difficult to achieve with straw alone. The cylinders, which weigh 5 kg, are moulded on an armature and have an observed life expectancy of 10-15 years.[230]

Veuille comments as follows about his cylindrical hive:

> Not being subject to the constraints of frames (encumbering mass of wood and artificially created gaps) the bees have total control of their living space which gives rise to the following:
> - the eight combs are built without interruption through the entire height of the hive according to the number of *ruchetons* and they rarely contain holes for transverse passages;
> - the combs are only fixed to the walls at spaced-out anchorage points; the interior traffic of the hive largely occurs via this peripheral clearance;
> - with the absence of internal corners there are no spaces prone to condensation, mould, development of pathogens, or invasion by parasites or predators;

- the visible results show good colony health as evidenced by the total disappearance of mycoses and a remarkably early development of the colonies with the first fine days in March.

Fig. 10.8: Cylindrical hive body in plaster and straw

Top-bars and spales

Fixing top-bars into boxes in the way Warré describes makes it slightly more difficult to remove combs should this ever be necessary. One solution is to put slots in the ends of the top-bars and locate them on pins in the rebate as shown in the illustration. Another solution is to use matchsticks set in holes in the rebate and top-bar. These can be broken off to ease cleaning the rebate, and the holes re-drilled. Other beekeepers have set the top-bars in a drop of beeswax or wedged them in place with matchsticks inserted between them and the vertical part of the rebate.

Fig. 10.9: Top-bars with slots and pins. Photo: Larry Garrett

Some Warré beekeepers are experimenting with nine instead of eight top-bars per box. The motives for this are that feral colonies often show comb spacing between centres (midribs) of as little as 30 mm and bees building small-cells of less than 5 mm tend to have closer comb spacing. To ensure that the bees build nine combs, wax starter strips on the bars are essential.

There have been occasional comments that the bees in some Warré hives are apparently reluctant to move down into the next box. I have certainly noticed considerable variation in the speed at which colonies do this and have generally put it down to variations in the quality of the queen; the health and genetics of the colony; the nectar flow; weather conditions; time of the season and the status of the existing nest etc. I do not see it as a problem, and certainly Warré does not mention it. However, he does make a comment that is relevant to this issue:

> This arrangement gives a space of 13 mm between the boxes. These 13 mm comprise the 9 mm of the top-bars and the 4 mm gap left by the bees at the bottoms of the combs, being the thickness of the bee body, for the bee, when working with its underside in the air, cannot extend the comb where its body is. This gap suits the bees in winter as it facilitates communication within the cluster of bees. If the gap did not exist, the bees themselves would create holes across the combs as they do in the frames of other hives. However, I consider the gap a fault, because the bees have to heat it almost wastefully in spring. It is a single fault, and moreover small beside the advantages that result from this arrangement, an even smaller fault than that of modern hives where the bees have uselessly to heat very much greater voids.[231]

Here he is concerned about the lowered thermal efficiency caused by splitting what was during the evolution of his hive a 400 mm deep comb into two 200 mm deep combs. But there is also the factor that the row of top-bars below the combs may be perceived by the bees as a kind of false floor that inhibits, at least for a while, the further development of comb downwards. Several Warré precursors, including John Gedde (1647-1697), instead of using top-bars, covered the tops of their hive elements with a board with a hole in the middle.[232] In Gedde's case it was 100 mm square. This allowed central combs to grow through into the box below and maintain continuity of the nest. To remove a box, the hole was closed with a sliding shutter which would have cut through any comb bridging the boxes.

One possible way of minimising any inhibition that may exist is to orientate the top-bars so that their widest side is in a vertical plane. Gilbert Veuille uses this orientation for his cylindrical 'Warré' modification.[233] It is also used by Viktor Shapkin, a natural beekeeper in Russia.

Fig. 10.10: Vertical top-bars in an octagonal hive box and in a log hive (inset).
Photos: Dietrich Vageler (Brazil) and Viktor Shapkin (Russia) respectively

Mounting the bars can be done by cutting angled slots in two opposite top rims of the box, with or without the 10 x 10 mm rebate present, angling the ends of the top bars to match, and resting them in the slots. This opens up the gaps between the top bars from 12 mm to 27 mm. Furthermore, the top-bar is more rigid in this orientation. The method was undergoing trials in Warré hives at the time of writing this book so it is not possible yet to give any reports on its performance.

The most radical departure from Warré's top-bars is to do without them altogether and just use spales, i.e. thin sticks supported in the form of a cross in the hive box. In skep beekeeping, these are pushed like skewers through the walls of the skep before populating it. Their function is to support the comb. When using spales in Warré hives, only the top box has a support all the way across its top. This may be just a board or the usual top-bars and a quilt. Whilst using spales alone allows the bees to make comb even more naturally than with top-bars in all boxes, it creates greater difficulties for the beekeeper because the boxes have to be cut off with a wire, for example a cheese-wire or potter's wire with handles. This is perfectly feasible provided that it is done with care. The wire is drawn along the length of the combs rather than against their sides. The process is assisted by supporting the weight of the box on thin wedges after prising the boxes up slightly with a hive tool. The method has been used for centuries in Japan in the management of *Apis cerana* colonies in boxes resembling those of Warré.

A management alternative to modifying the top-bars is to put a top-bar with comb on it near the middle of the new box that is about to be occupied. This can act as a bait to draw the bees down.

Fig. 10.11: Box with spales, harvesting from an *Apis cerana* colony in Japan.[234]
Photo: Syouichi Morimoto

Frames

The People's Hive with frames

In some countries the law requires combs to be easily movable on some sort of frame. What a frame comprises is not usually specified and presumably could simply be a top-bar that is not fixed with nails. But in practice, the expectation is that comb should be removable with the rough handling commonly applied to frames that encompass the whole perimeter of the comb.

Although, as we have seen, top-bar combs can be removed and replaced, more care is needed. Some inspectors may not take too kindly to having to go through an apiary full of top-bar hives. In which case, a compromise is available for the People's Hive. This is the People's Hive with frames. A translation into English of the relevant pages of the 5th edition of *Beekeeping for All* is available in which the construction of this version is described.[235] In it Warré wrote:

> Nowadays, I recommend without hesitation the People's Hive with fixed combs, even for very large enterprises. [...] However, out of respect for the freedom of my readers, I will describe the People's Hive in its three forms: fixed comb, ordinary frames, open frames with closed ends [i.e. semi-frames, *Tr.*].

This frame version retains other essential features of the People's Hive such as the top-bar cloth and quilt, as well as the management by nadiring. The top-bar cloth and quilt also help somewhat with the retention of the nest scent and heat. It just leaves the ends of the combs open because of the bee space between the frame side-bars and the walls. There is also the increased waste space, ecological footprint and cost caused by the frames, but this is offset in the versions of Marc Gatineau[236] and Gilles Denis[237] by using wood that is as thin as 7 mm in the frame construction.[238]

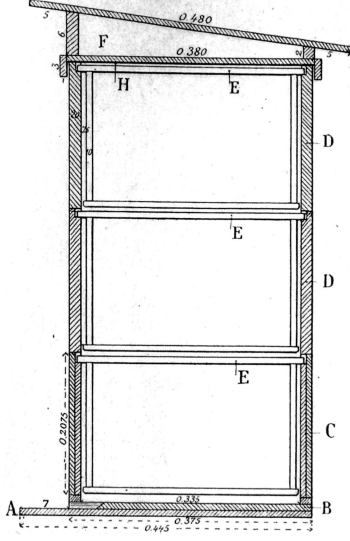

Fig. 10.12: The People's Hive with frames; also illustrates Warré's simplified roof[239]

Roger Delon's frame

Roger Delon developed a modified top-bar on which was fixed a 'U' shape of 3 mm stainless-steel wire that surrounds the remaining three edges of the comb.[240] His main reason for requiring better stability was his extensive use of migratory beekeeping in the Vosges and Jura, involving some 600 hives. The wire is 'invisible' to the bees to the extent that it eventually becomes embedded in the comb allowing the comb to be built past it to the walls. This safeguards the retention of nest heat of more natural comb. The only objections to this on the grounds of sustainability are that stainless-steel has a high ecological footprint and the construction of Delon frames increases the complexity and cost of the operation.

Fig. 10.13: Delon frame

Gilles Denis' semi-frame

Gilles Denis, also a practitioner of large-scale migratory beekeeping with his modified Warré hives, observed that in his top-bar hives the combs were generally stuck to the walls only along the top half of their edges. He therefore devised a semi-frame comprising a top-bar and two 90 mm side-bars. He reserves these for breeding and uses plain top-bars in his production hives.[241] As we have mentioned above, Warré also conceived a frame type without a bottom-bar.

Although semi-frames use a renewable material, namely wood, they do not completely accord with Thür's criterion of retention of nest scent and heat as the combs are prevented from sealing to the walls at the edges, because there is a bee space between the side-bars and the hive walls. However, semi-frames score higher in terms of sustainable use of resources compared with Delon steel frames.

Fig. 10.14: Semi-frame for Warré hive. Photo: Gilles Denis

Whilst some form of enhancement to the top-bars makes inspection of the Warré hive easier, it should not be overlooked that an apiary with, for example, only three hives could have a dozen or more brood boxes for the bee disease inspector to go through. It may make more economic sense if disease control in such hives is expedited by sampling bees from the comb edge and submitting them for analysis.[242]

Top-bar cloth, quilt and roof

As an alternative to hessian or burlap, some beekeepers use fly screen mesh on the top-bars.[243] This needs to be robust enough so as not to be frayed by the bees and the holes need to be fine enough to prevent the passage of bees. In practice, something less than eight mesh (2-3 mm gaps) is used. The bees normally propolise the exposed undersides of the cloth, covering all the small holes in it. However, both Warré[244] and Guillaume[245] observe that the bees sometimes unpropolise the holes in the covering to improve local ventilation. I have not yet seen evidence of this on any of the cloths that I have removed. Just occasionally a colony chews a few small holes in a top-bar cloth, though not into the quilt. Whether this is to improve ventilation, or because of inadequate flour paste treatment, is unclear.

Modifications of the quilt have comprised doing away with it altogether as in Roger Delon's 'Stable Climate Hive'. Over the top bar cloth are two boards one above the other which trap a closed air space. The overall thickness of the unit is 30 mm. Delon used a top-bar cloth.

Several Warré beekeepers do not use Warré's double-pitched roof. One reason for this is to make the hives more transportable. Warré's roof has two design principles which are not generally incorporated in other hive roofs. One is the very well ventilated space under the outer boards that helps protect it from the heat of the sun. The other is the use of wood instead of metal to minimise the disturbance to the bees from the drumming of rain or hail.

A number of newcomers to the Warré hive, including myself, were surprised to find that the ventilated space does not communicate directly with the quilt. Indeed, the hive is not designed to create a chimney effect, i.e. an air flow through the quilt. If there is any moisture or air movement from the hive out of the quilt it is only through cracks under the mouse board in the roof. In order to increase ventilation, or at least vapour transfer, some beekeepers have made small holes in the mouse board or used peg board, i.e. hardboard with small holes in it.

There has been some discussion by beekeepers in sub-arctic regions as to whether condensation moisture may build up in Warré's quilt, freeze, and drip on the bees in the spring thaw. This problem has not yet been reported as far as I am aware. Furthermore, I have checked all my quilts after the coldest winter months and on one occasion only, out of fifteen quilts, two had very slight dampness in the top 10-15 mm of the wood shavings filling. Deeper in the shavings, it was perfectly dry. This has been confirmed by others.

Frères and Guillaume have published an ingenious revision of Warré's arrangement at the top of the hive.[246] They dispense with Warré's quilt and instead use a single pitched roof which is deep enough to accommodate a 50 mm slab of expanded polystyrene and a contact feeder. The polystyrene is fixed to the upper surface of a sheet of hessian (burlap) sacking which in turn is tacked to the inside wall of the roof so as to allow the insulation to move up and down inside the roof. When there is no feeder present it rests on the top-bars. When a feeder is put in position before replacing the roof the hessian and insulation are pushed upwards. This ensures that insulation is in place even when feeding. The roof above the insulation is well ventilated on all four sides. The arrangement is used by Warré beekeepers in France but it has yet to be taken up in other countries.

In a revised version of the hive covering, Frères and Guillaume have substituted insulation made of biofibres, i.e. renewables, for the polystyrene.

Stand

In Warré's hive design, the legs/feet of the hive, made either of cast iron or wood, are fixed to the floor.[247] Both types are designed to spread the hive load outside the footprint of the floor itself and are wide enough to prevent the hive sinking into the ground. Some designs I have seen have ignored these features and just fixed simple wooden legs to the floor, resulting, to my knowledge, in at least one hive falling over.

Warré also preferred to have the hive very close to the ground with the entrance 100-150 mm above it. He argued that it would be less subject to temperature fluctuations and gusts of wind. Furthermore, tired foragers resting on the ground would be able to get back into the hive relatively easily, especially in windy conditions.

I use separate 300 mm high stands for my hives as shown in the illustration. I have not noticed that it created any difficulties for returning foragers. Furthermore, all four legs are outside the footprint of the hive so it is very stable – an essential feature in the windy, maritime climate around my apiary. Sustainable use of resources is maximised by constructing the stand from scrap timber. To ease levelling and increase stability, I put a recycled paving slab of at least 450 mm square under each stand.

A simple alternative is to raise the hive at least 150 mm from the ground using building blocks of a suitable size and preferably recycled.

Fig. 10.15: Warré hive stand

Removing combs on top-bars

Most Warré beekeepers prefer not to remove combs from boxes at all. However, we have already considered in this chapter making the compromise of using frames or semi-frames to ease comb removal for inspection. Here we look at how to remove combs on top-bars. Firstly, unless the combs have been formed parallel to one

10.16: Hooked knife for removing combs on top-bars,
full view and close-up of blade (inset). Photos: Bill Wood

another and not crossed between top bars, it is impossible to remove them without doing a lot of damage. The comb is usually fixed to the walls at each end at least half way down from the top of the box. This attachment has to be cut with a knife. Although it could be done from underneath by tilting the box on one side, an easier solution is to do it from above using a hooked knife designed by Bill Wood (see illustrations).

The 16 gauge (circa 1.25 mm) stainless steel knife blade is ¼ inch (6 mm) wide by between 1⅛ (28 mm) and 1¼ inch (32 mm) long, and is welded to a ¼ inch (6 mm) diameter stainless steel shaft, or is formed by bending the shaft through a right-angle and grinding the blade to shape. In the latter case, provided that the bend has not weakened the steel, a narrower blade could be made.

It is used from the top of a box to free comb-to-box attachments in order to inspect an individual comb taken from a populated hive.

The knife is inserted downwards between chosen combs with the shaft touching the inside of the box and the blade pointing away from the box wall. Insertion is stopped at a marked distance where the knife is very slightly above the top bars of the box beneath. Then the knife is rotated 90 degrees and pulled upwards, cutting close to the box side and freeing the comb. It is important that the cutting is always towards the top-bar that the comb is on, avoiding any downward force on the comb that might lead to detaching it from the bar.

When the knife reaches the underside of the top bar, the knife is further rotated, moving the end of the shaft away from the box as the blade is angled parallel to the comb. The knife is then extracted with the blade now pointing in the opposite direction to that when it was inserted. The blade undergoes a 180 degree rotation in the process of cutting one end of a comb.

Bill Wood reports that this procedure only minimally disrupts comb cells. In cutting free and inspecting two full combs, less than 2 ml (< ½ teaspoon) of honey from crushed cells spilled down to the bottom board.

Fig. 10.17: Warré comb removed for inspection. Photo: Karman Csaba.

Intensely melliferous localities

Anomalous behaviour of the Warré hive has been noted where there is very intense nectar flow over most of the foraging season. In Alberta, Canada, there are large areas of high-yielding legume crops such as alfalfa, sweet clover, alsike and white Dutch clover. These, coupled with the ubiquitous oilseed rape (canola), make for artificially high nectar availability to bees over long periods. It is out of proportion with the flow which would occur with more natural flora. A hint that things were not quite as Warré describes in his book came during the 2008 season and was confirmed in 2009. The bees filled up to six boxes with comb very quickly. Not only that, but the brood nest became extended over most of the boxes. Winter losses were only about 10% in 2008/9 but rose to about 75% in 2009/10.

Dissection of the nests revealed that the cluster had been confined to a narrow column of honey-free comb up the middle of the hive surrounded by a cylinder of honey. It was therefore surmised that nectar was coming in so fast that the bees were depositing more honey than usual in the outer perimeter of the brood nest, thus restricting the space available for the queen to lay. This resulted in a honey-bound brood area and consequently a columnar cluster shape that was thermally inefficient for wintering.

A way of dealing with this problem under consideration at the time of writing involves widening the internal area of the boxes and the careful application of supering. (See the Addendum for an update)

APPENDIX 1

Table of thermal conductivities of hive materials (W/m.K)

Air	0.024-0.026
Empty comb with a layer of bees	0.027[248]
Wood shavings (loosely or densely packed)	0.04-0.08
Cork	0.043
Sawdust	0.08-0.09
Packed straw	0.054-0.15
Softwood	0.10-0.14
Hardwood	0.15-0.17
Plaster (gypsum)	0.48
Glass	0.63
Earthenware	1.31
Steel (stainless) ·	16
Steel (carbon, 1%)	43

APPENDIX 2

Plans for constructing a Warré hive – The People's Hive

The plans shown below are based as closely as possible on those in the 12th edition of *Beekeeping for All* published by Warré (1948). The page numbers in the text refer to the pages in the book.

At the end of this document are included plans for constructing a box with a window according to the modification of Warré's hive introduced by Frèrès & Guillaume (1997). We include this variant of the box because many beekeepers, especially beginners, find it helpful for monitoring the progress of colony development. However, we warn that adding windows increases the complexity and expense, not to mention the consumption of resources.

The plans shown here are based on boxes of 20 mm thick wood, the minimum that Warré regarded as sufficient (p. 52). However, he recommended 24 mm for improved rigidity. Thicknesses of 38 and 50 mm have also been used in colder climates. Any change to the box wall thickness should ideally retain the internal measurements of 300 x 300 x 210 mm and will thus require all other components of the hive apart from the legs to be re-sized.

The hive

There is no direct passage of air from the top box to the vents under the roof. The quilt is filled with a suitable insulating material of plant origin such as chopped straw, wood shavings, dried leaves etc. As well as a top-bar cover cloth there is a cloth fixed to the bottom of the quilt to retain its contents. A suitable material for these cloths is hessian sacking (burlap).

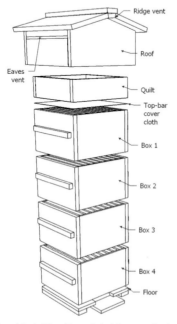

Fig. A2.1: The People's Hive exploded

Fig. A2.2: The People's Hive

The boxes

The basic box is butt-jointed at the corners. Nailing with seven galvanised nails 65 x 2.65 mm makes a strong joint. Four screws of the same length also suffice. Glue is unnecessary. Obviously half-joints or finger joints make a stronger box, although this increases the complexity. Any cracks are filled by the bees. However, it helps if the boxes sit on each other without rocking significantly. Use of a square throughout assembly ensures this. If in doubt, knock nails in only partially at first.

Warré does not specify the size of the handles, but 300 x 20 x 20 mm bars, fixed with three nails and glued have been found to work well in practice, including when lifting the entire hive with the bottom box handles. The upper edge of the bars may be sloped to shed rainwater.

The spacing between all bars and between the end bars and the walls is 12 mm. The rebates can be cut with a table saw, a rebating plane, or even with a hand saw provided that a guide is affixed to the wood first. Warré specifies a top-bar length of 315 mm. If it is extended close to 320 mm, then it better spans the rebate and leaves no cavity for pests to occupy.

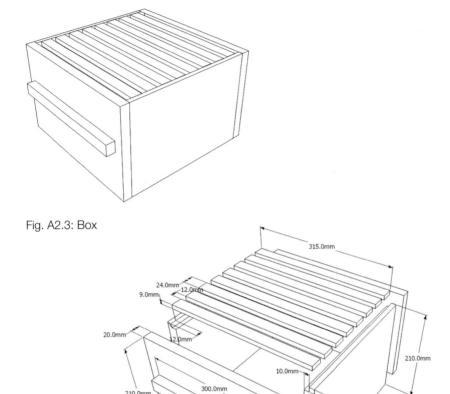

Fig. A2.3: Box

Fig. A2.4: Box exploded

The quilt

The thickness of the walls of the quilt could be as little as 10 mm as it supports only the roof. The quilt shown here is the same size as the box. Warré suggested making the quilt 5 mm smaller than the box to allow the hessian/burlap quilt contents retainer to be brought up the sides of the quilt and fixed there.

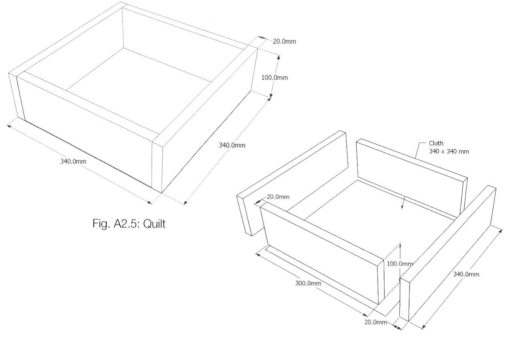

Fig. A2.5: Quilt

Fig. A2.6: Quilt exploded

The roof

This is the more complicated of the two roof designs that Warré presents in *Beekeeping for All*. It has a ventilated cavity immediately under the upper surfaces of the roof. This is to dissipate solar heat. Inside is a cover board (or 'mouse board') which rests on the quilt. Thus there is no communication from the quilt to the ventilated roof cavity.

Fig. A2.7: Roof

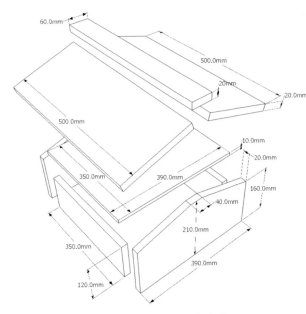

Fig. A2.8: Roof exploded

The example given here uses 20 mm wood for the outer structure. This is based on Warré's plans. Wood of 15 mm thickness would be perfectly adequate and result in a lighter roof. There is 10 mm play between the roof inner walls and the quilt outer walls to ease placement and removal of the roof.

If thinner wood is used for the inclined boards, they may be nailed to the ridge board from their undersides by supporting the three pieces inverted, with blocks suitably positioned to give approximately the correct angle. This unit of three boards is then nailed on to the gables. Nailing the sloping boards to the ridge helps form a seal against driving rain, especially if the sloping boards are thinner and thus inclined to warp.

Fig. A2.10 shows a cut-away view of the quilt in the roof. Note that the lower rim of the roof projects below the junction in which reside the top-bar cloth and the cloth fixed to the bottom of the quilt. This prevents rainwater from running into that joint and wicking into the top-bar cloth and quilt.

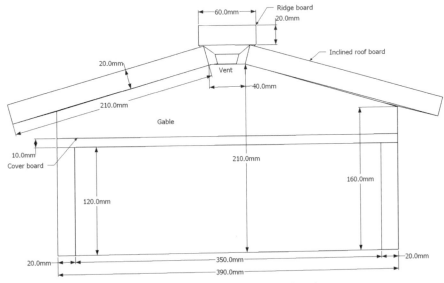

Fig. A2.9: Gable elevation of roof

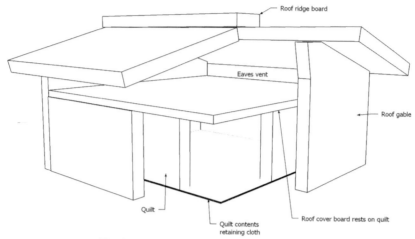

Fig. A2.10: Cut-away view of roof and quilt

The floor

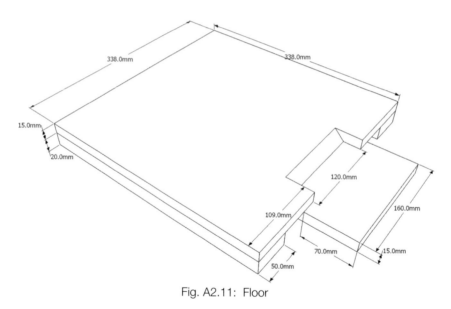

Fig. A2.11: Floor

Warré recommended 15-20 mm for the thickness of the floor. He gave no guidance on the thickness of the battens underneath it.

The notch in the floor serves as the entrance and is 40 mm deep for hive walls of 20 mm thickness. If required, the 160 mm wide alighting board can extend right to the back of the floor to give the latter added rigidity.

Note that the floor is 2 mm narrower in both horizontal directions compared with a box. Warré suggested allowing 1 mm all round to promote drainage of rainwater.

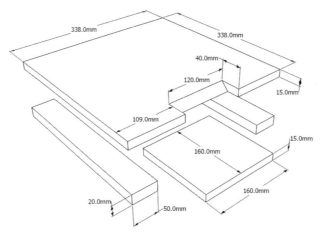

338.0mm
338.0mm
40.0mm
120.0mm
15.0mm
109.0mm
15.0mm
160.0mm
160.0mm
20.0mm
50.0mm

Fig. A2.12: Floor exploded view

The legs

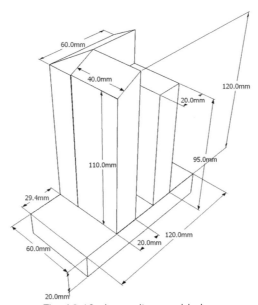

60.0mm
40.0mm
120.0mm
20.0mm
95.0mm
110.0mm
29.4mm
120.0mm
20.0mm
60.0mm
20.0mm

Fig. A2.13: Leg unit assembled

The leg has a wide foot attached to prevent the hive from sinking into the ground and possibly toppling. Furthermore it gives a 20 or 30 mm projection outside the footprint of the floor, thus greatly increasing stability.

The leg places the hive entrance relatively close to the ground compared with legs in common beekeeping practice. However, Warré regarded a low entrance as important (pages 46-48).

In the above drawing, the right hand pillar of the leg has the underside of the floor resting on it. The angled left hand pillar is nailed from both sides to the side of the floor with a total of four nails. The difference in height of the right and left pillars is 25 mm, which will work for the floor corner overall thickness of 35 mm shown above. If the floor corner is less than 25 mm thick, then the right pillar should be lengthened accordingly. The feet may be arranged in a pinwheel (Catherine-wheel) arrangement or constructed as two mirror symmetrical pairs and placed in mirror symmetry accordingly.

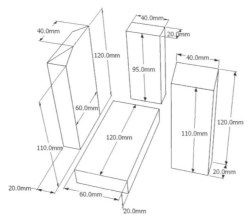

Fig. A2.14: Leg unit exploded

Box with window and shutter

This is based on the design by Frères & Guillaume (1997).

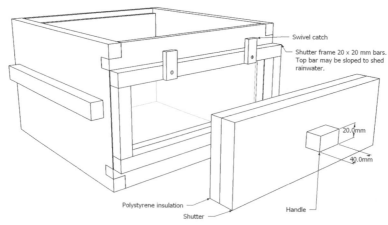

Fig. A2.15: Box with window and shutter

This design places the insulation flush against the glass, the depth of the cavity exactly matching the depth of the shutter. To minimise jamming of the shutter in wet conditions, it may be made a millimetre or more smaller in each direction.

In this version, the window frame bars are jointed to the box walls with mortise and tenon joints. Alternatively, butt joints with three nails or two screws at each joint should give satisfactory service.

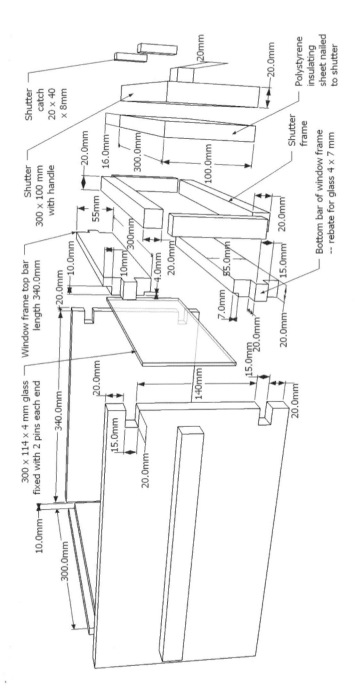

Fig. A2.16: Box with window and shutter exploded

APPENDIX 3
PLANS FOR CONSTRUCTING A WARRÉ HIVE LIFT

As Warré hives are managed by nadiring, once a hive has grown beyond two boxes it is difficult to lift it without an assistant. As each box has sturdy, projecting handles, it was appropriate to design a fork lift that picks up the hive by its handles. The lift described here, the first of which is believed to have been produced by Marc Gatineau, works on the principle of the French guillotine. A board slides vertically in two grooves and is lifted with a windlass. The groove and the board edges are thoroughly rubbed with beeswax before assembly. This is to promote a smooth action. Two fork tines are mounted on the board and the hive box fits between them. The hive can be picked up by the bottom box or by either of the two above it, Fig. 9.10.

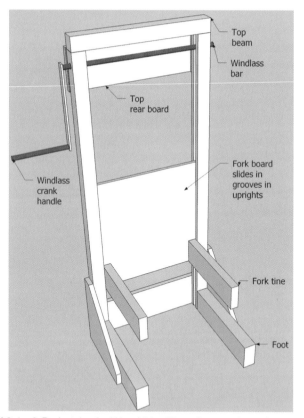

Fig. A3.1: 3-D sketch of a Warré hive lift, pulleys and cord not shown

The fixing of the fork tines to the rear board requires careful attention. On the bottom of the back of the 13 mm thick fork board is glued centrally and flush with the bottom edge another plywood board measuring 412 mm x 200 mm x 13 mm.

This is to reinforce the fork board in the region of the tines.

The tines of the fork are cut perfectly square at the ends, butt-jointed with glue onto the board and clamped in place at right angles with the help of supporting blocks. When the glue is dry, two 100 mm wood screws are screwed into the tines through holes drilled and countersunk into the back. The screws are size 12, i.e. 6 mm thick at the widest end. The screws are lubricated with wood glue when inserting so that any play is taken up by the glue as it sets. The extra board glued on the back takes the vertically sliding unit ('guillotine blade') to a thickness of 26 mm and confers greater rigidity. It must of course clear the inner edges of the uprights. The clearance here is 1 mm each side.

Other successful methods of fixing the tines include sturdy shelf brackets (Steve Ham, Spain) and tee hinges (Bill Wood, USA).

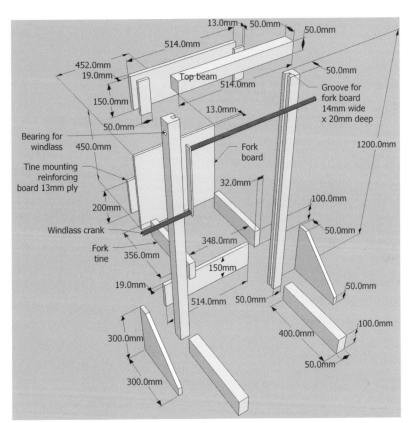

Fig. A3.2: 3-D exploded view of a Warré hive lift

The top pulley is tied to the top beam and the bottom pulley to the fork board. The lifting cord is tied to the windlass axle, threaded through the two double pulleys and tied to the fork board (see photo of lift, Fig. 9.10). Pulleys are not essential, but they do ease lifting of a tall, honey-laden hive and allow movement to be stopped at any point without securing the crank. The top rear board is set out on 13 mm spacers to clear the windlass cord and top pulley.

The windlass and crank handle could be made of steel pipe used for electrical conduit etc. or reclaimed from other machinery such as cars, mangles, honey extractors etc. To increase durability of the windlass bearings, they may be lined with steel bushes and greased.

No effort has been made to reduce the weight of the lift, which is somewhat heavy to carry, but several dimensions could be reduced by a few millimetres without significant loss of rigidity, for example the thickness of the upright, top beam and feet as well as of the rear and triangular bracing boards. The fork board should not be thinner than 13 mm.

Useful accessories incorporated in a Warré lift include wheels and a hive weighing scale. The ability to adjust the gap between the fork tines would be a helpful development for picking up hives with boxes made with wood thicker than 20 mm. A range of lift designs is illustrated on the Warré beekeeping English web portal.[249]

APPENDIX 4
A SELECTION OF INTERNET RESOURCES

Bee-friendly beekeeping: http://www.bee-friendly.co.uk

Warré English web portal: http://warre.biobees.com

Beekeeping for All: http://warre.biobees.com/bfa.htm

Yahoo e-group: http://uk.groups.yahoo.com/group/warrebeekeeping

David Heaf's beekeeping pages: http://www.dheaf.plus.com/warrebeekeeping/beeindex.htm

Friends of the Bees: http://www.friendsofthebees.org

The Natural Beekeeping Trust: http://www.naturalbeekeepingtrust.org

Bees for Development: http://www.beesfordevelopment.org/

World Save the Bee Fund: http://www.save-bee.com/en/

ACKNOWLEDGEMENTS

I thank George Burgess for suggesting that I take up beekeeping and for his generous and friendly mentoring when I started. I thank Dr. Johannes Wirz (Goetheanum, Switzerland) for introducing me to Warré beekeeping, Bernhard Heuvel (Germany) for his helpful advice on sustainable beekeeping and Phil Chandler (Devon) and Dr William Hughes (Wales) for kindly offering web space to support this work. I also thank John Phipps, editor of 'The Beekeepers Quarterly', for suggesting the series of articles that gave rise to this book, for his helpful comments towards them, and for enabling the Warré approach to be made known in the anglophone regions. Embarking on the attempt to write this book at all I owe to suggestions at various times from John Phipps, Jeremy Burbidge (Northern Bee Books), Maddy Harland (Permaculture Magazine) in that temporal order. Maddy's prompt was the one that tipped the balance in favour of my making a start.

For her encouragement and helpful discussions during the writing of the book, and professional help with editing the manuscript, I thank my wife Pat, who was also co-translator of Émile Warré's *Beekeeping for All*. Erik Berrevoets, Jérôme Alphonse and Heidi Hermann kindly commented on the content of the text, and the detailed scrutiny of the manuscript by John and Chris Haverson greatly enhanced it. I warmly thank all five for their help. Any errors that may remain are entirely my responsibility.

I thank Dr. Naomi Saville for her helpful advice on appropriate hives design in Jumla, Nepal. Last, but in no way least, I thank the enormous input I have received from beekeepers who are mostly members of the Yahoo Warré English e-Group. I have acknowledged some of their specific contributions at various places in the text, especially regarding many of the photos of Warré beekeeping.

GLOSSARY

Acaricide Chemical for killing mites.

Bailey comb change Moving a colony onto new comb in a single operation. A box usually with frames of foundation is placed over the colony. When some comb is built the queen is found and placed on it. A queen excluder is placed below her box over the old box. After three weeks, the old box is removed.

Bole A hollow in a wall for sheltering one or more skeps.

Cleam To daub a skep with cow dung etc.

Eke A shallow hive body element that can go above or below the brood to create space for any purpose.

Genome The totality of the genes of a particular organism.

Hackle A conical shaped straw roof used on skeps.

Immune pathway redundancy The presence of unused but potentially usable routes or pathways at the molecular biological level for developing an immune response.

Introgression In this context refers to the movement of genes from elsewhere into the bee breed of interest.

Midrib The middle wall between the two sides of the comb forming the bases of the cells.

Nadir To add a hive element underneath an occupied hive. The element that is added underneath.

Nuc/Nucleus A small bee colony usually on five or six frames with brood of all ages, stores and a marked queen.

Patriline A group of full sisters in a colony from the same drone (father).

Skep A round hive of straw, reeds, or other flexible plant material, often coated with cow dung which may also be mixed with clay.

Spale Thin cross-stick inserted through skep or other hive as comb support.

REFERENCES

Alaux, C., Brunet, J-L., Dussaubat, C., Mondet, F., Tchamitchan, S., Cousin, M., Brillard, J., Baldy, A., Belzunces, L. P. & Le Conte, Y. (2009) Interactions between *Nosema* microspores and a neonicotinoid weaken honeybees (*Apis mellifera*) *Environmental Microbiology* **12**(3) 774-782.

Alaux, C., Ducloz, F., Crauser, D. & Le Conte, Y. (2010) Diet effects on honeybee immunocompetence *Biol. Lett.* Published online doi:10.1098/rsbl.2009.0986.

Atienza, J., Jiménez, J. J., Bernal, J. L. & Martin, M. T. (1993) Supercritical Fluid Extraction of Fluvalinate Residues in Honey. *J. Chromatogr.* **655**, 95-99.

Bailey, L. & Perry, J. N. (2001) The natural control of the tracheal mite of honey bees. *Exp. Appl. Acarol.* **25**, 745-749.

Baum, K. A., Rubink, W.L., Pinto, M. A. & Coulson, R.N. (2005) Spatial and Temporal Distribution and Nest Site Characteristics of Feral Honey Bee (Hymenoptera: Apidae) Colonies in a Coastal Prairie Landscape *Environ. Entomol.* **34**(3), 610-618.

Becher, M. A., Scharpenberg, H. & Moritz, R. F. A. (2009) Pupal developmental temperature and behavioral specialization of honeybee workers (*Apis mellifera* L.). *J Comp Physiol A* **195**, 673-679.

Berrevoets, E. (2009) *Wisdom of the bees - principles of biodynamic beekeeping*. Anthroposophic Press, NY.

Berry, J. (2009) Pesticides, Bees And Wax - Measuring the effects of the compounds we use. *Bee Culture* **137**(1), www.beeculture.com/storycms/index.cfm?cat=Story&recordID=626.

Berry, J. A. & Delaplane K. S. (2001) Effects of comb age on honey bee colony growth and brood survivorship *J. Apicultural Res.* **40**(1), 117-122.

Berry, J. A., Owens, W. B. & Delaplane, K. S. (2010) Small-cell comb foundation does not impede Varroa mite population growth in honey bee colonies. *Apidologie* **41**, 40-44.

Bujok, B., Kleinhenz, M., Fuchs, S. & Tautz, J. (2002) Hot spots in the bee hive. *Naturwissenschaften* **89**, 299-301.

Bujok, B. (2005) *Thermoregulation im Brutbereich der Honigbiene* Apis mellifera carnica. Thesis. (www.opus-bayern.de/uni-wuerzburg/volltexte/2005/1590). Theodor-Boveri-Institut für Biowissen-schaften, Fakultät für Biologie.

Chandler, P. J. (2007) *The Barefoot Beekeeper.* Self published, available at www.biobees.com.

Chapleau, J. P. (2003) *Experimentation of an Anti-Varroa Screened Bottom Board in the Context of Developing an Integrated Pest Management Strategy for Varroa Infested Honeybees in the Province of Quebec.*http://www.apinovar.com/articles/AV-BOTTOM%20BOARD.pdf

Cheshire, F. R. (1888) *Bees and beekeeping; scientific and practical. Vol. II - Practical.* The Bazaar Exchange and Mart Office, London.

Christ, J. L. (1783) *Anweisung zur nützlichen and angenehmen Bienenzucht für alle Gegenden, Frankfurt & Leipzig.* Transcribed and republished by Eric Zeissloff, Pfulgresheim, 2006. Ordering information: z.eric@onlinehome.de. English notes on the essential information about Christshive in this book can be downloaded as a PDF document at http://warre.biobees.com/christ.htm.

Coffey, M. F., Breen, J., Brown, M. J. F. & McMullan, J. B. (2010) Brood-cell size has no influence on the population dynamics of Varroa destructor mites in the native western honey bee, *Apis mellifera mellifera. Apidologie* DOI: 10.1051/apido/2010003. apidologie.org.

Conrad, R. (2007) *Natural beekeeping - organic approaches to modern apiculture.* Chelsea Green Publishing, Vermont.

De Jong, D. (1982) Orientation of comb building by honeybees. *J. Comp. Physiol. A.* **147**(4), 495-501.

Denis, G. (2008) *Mode d'emploi de la ruche Warré* (How to use the Warré hive), Published by the author: 2 Rue Jean Monnet, 42650 St Jean Bonnefonds, France (tel. 33 (0)4 53 86 23). ISBN 978-2-9533201. www.ruche-warre.com.

Dewhurst, R.D. (2006) Biting Back: The behavioural adaptation of the dark European honeybee to varroa and the conservation of the native bees of Cornwall. *Beekeepers Quarterly* **86**, 15-18.

Ellis, A. M., Hayes G. W. & Ellis J. D. (2009) The efficacy of small cell foundation as a varroa mite (*Varroa destructor*) control *Exp. Appl. Acarol.* **47**, 311-316.

Ellis, A. M., Hayes, G. W. & Ellis, J. D. (2009b) The efficacy of dusting honey bee colonies with powdered sugar to reduce varroa mite populations. *J. Apicultural Res.* **48**(1), 72-76.

Evans, J. D. (2006) Pathway and transcriptional insights into honey bee immunity from the Honey Bee Genome Project. The Fifth International Symposium on Molecular Insect Science, May 20-24, 2006, Tucson, Arizona USA. www.insectscience.org/6.46/.

Forsgren, E., Olofsson, T. C., Vásquez, A. & Fries, I. (2009) Novel lactic acid bacteria inhibiting *Paenibacillus larvae larvae* in honey bee larvae. *Apidologie* online DOI: 10.1051/apido/2009065.

Frazier, M., Mullin, C., Frazier, J. & Ashcraft, S. (2008) What have pesticides got to do with it? *Am. Bee. J.* **148**, 521-523.

Free, J. B. & Williams, I. H. (1974) Factors determining food storage and brood rearing in honeybee (*Apis mellifera* L.) comb. *J. Entomol. Series A* **49**, 47-63.

Frères, J-M. & Guillaume, J-C (1997) *L'Apiculture Écologique de A à Z*. Guillaume J-C (ed. & publ.) Villelongue-Dels-Monts.

Fries, I. & Bommarco, R. (2007) Possible host-parasite adaptations in honey bees infested by Varroa destructor mites. *Apidologie* **38**, 525-533.

Fries, I. & Camazine, S. (2001) Implications of horizontal and vertical pathogen transmission for honey bee epidemiology. *Apidologie* **32**, 199-214.

Fries, I. Imdorf, A. & Rosenkranz, P. (2006) Survival of mite infested (*Varroa destructor*) honey bee (*Apis mellifera*) colonies in a Nordic climate. *Apidologie* **37**, 564-570.

Gatineau, M. (2007) *L'apiculture, telle que je l'aime et la pratique* (The beekeeping that I love and practise) Self published, available at www.apiculturegatineau.fr.

Gedde, J. *A new discovery of an excellent method of beehouses, and colonies, which frees the owners from the great charge and trouble that attends the swarming of bees, and delivers the bees from the evil reward of ruine, for the benefit they brought their masters* (London, Printed for the author, 1677. 3rd edition, enlarged, with several objections answered.) A transcript of the part of this concerned with the Gedde hive is downloadable as a PDF at warre.biobees.com/gedde_transcript.pdf. A scan of Gedde's book is downloadable at www.culturaapicola.com.ar/apuntes/libros/468_Gedde.pdf.

Gillette, C. P. (1900) Effects of Artificial Foundation on the Building of Honey Comb. In: *Proceedings of the 21st Annual Meeting of the Society for the Promotion of Agricultural Science*.

Gillard, M., Charriere, J.D., & Belloy, L. (2008) Distribution of *Paenibacillus larvae* spores inside honey bee colonies and its relevance for diagnosis. *J. Invertebrate Pathology* **99**(1), 92-95.

Gilliam, M. (1997) Mini review: Identification and roles of non-pathogenic microflora associated with honey bees. *FEMS Microbiology Letters* **155**, 1-10.

Goodwin, R. M., Ten Houten, A. & Perry, J. H. (1994) Incidence of American foulbrood infections in feral honey bee colonies in New Zealand. *NZ J. Zool.* **21**, 285-287.

Graham, S., Myerscough, M. R., Jones, J. C. & Oldroyd, B. P. (2006) Modelling the role of intracolonial genetic diversity on regulation of brood temperature in honey bee (*Apis mellifera* L.) colonies. *Insectes Sociaux* **53**, 226-232.

Groh, C., Tautz, J. & Rössler, W. (2004) Synaptic organization in the adult honey bee brain is influenced by brood-temperature control during pupal development. *PNAS* **101**(12), 4268-4273.

Grozinger, C. (2010) The effect of reproduction on honey bee queen physiology, behavior and pheromone production. For a mini review see http://ento.psu.edu/directory/cmg25.

Harbo, J. R. & Harris, J. W. (2004) Effect of screen floors on populations of honey bees and parasitic mites (*Varroa destructor*) *J. Apicultural Res.* **43**(3), 114-117.

Harrison J.M. (1987) Roles of individual honeybee workers and drones in colonial thermogenesis. *J.Exp.* Biol. **129**, 53-61.

Hauk, G. (2002) *Toward Saving the Honeybee*. Biodynamic Farming and Gardening Association, San Francisco.

Heaf, D. J. (2011) Do small cells help bees cope with Varroa? *The Beekeepers Quarterly*, **104**, 39-45.

Human, H., Nicolson, S. W. & Dietmann, V. (2006) Do honeybees, Apis mellifera scutellata, regulate humidity in their nest? *Naturwissenschaften* **93**, 397-401.

Imperial College Consultants Limited (2008) *Honeybee health (risks) in England and Wales*. A report to the National Audit Office by Imperial College Consultants Limited. September 2008. www.nao.org.uk//idoc.ashx?docId=AC47B21D-2085-47EF-BF51-39904B555F72&version=-1.

Jaffé, R., Shaibi, T., Dietemann, V., Kraus, F.B., Crewe, R. & Moritz, R. F. A. (2007) Comparing honeybee densities: European semi-natural habitats versus African deserts (poster). Sustainable Neighbourhood - from Lisbon to Leipzig through Research (L2L): May 8th - 10th, Leipzig, Germany.

Johnson, R. M., Ellis, M. D., Mullin, C. A. & Frazier, M. (2010) Pesticides and honey bee toxicity – U.S.A. *Apidologie* **41**, 312-331.

Jones, J. C., Myerscough, M. R., Graham, S. & Oldroyd, B. P. (2004) Honey bee nest thermoregulation: diversity promotes stability. *Science* **305**, 402-404. (doi:10.1126/science.1096340).

Jones, J., Helliwell, P., Beekman, M., Maleszka, R. J. & Oldroyd, B. P. (2005) The effects of rearing temperature on developmental stability and learning and memory in the honey bee, *Apis mellifera*. *J. Comp. Physiol. A* **191**, 1121-1129.

Keeling, C. I., Slessor, K. N. Higo, H. A.& Winston, M. L. (2003) New components of the honey bee (*Apis mellifera* L.) queen retinue pheromone. *Proc. Natl. Acad. Sci. U S A*. **100**(8), 4486-4491.

Kefuss J., Vanpoucke J., Bolt M. & Kefuss C. (2009). Practical Varroa Resistance Selection for Beekeepers. In: *Scientific Program of the 41st Apimondia Congress*, Montpellier, p.82 (see www.apimondia.org)

Keller, I., Fluri, P. & Imdorf, A. (2005) Pollen nutrition and colony development in honey bees: Part I. *Bee World* **86**(1), 3-10.

Kocher, S. D., Richard, F-J., Tarpy, D. R. & Grozinger, C. M. (2009) Queen reproductive state modulates pheromone production and queen-worker interactions in honeybees. *Behavioral Ecology* **20**(5) 1007-1014.

Kockelkoren, P. (1995) Ethical Aspects of Plant Biotechnology - Report to the Dutch Government Commission on Ethical Aspects of Biotechnology in Plants, In: *Agriculture and Spirituality - Essays from the Crossroads Conference at Wageningen Agricultural University (Appendix 1)*. International Books, Utrecht. 99-105.

Köpler, K., Vorwohl, G. & Koeniger, N. (2007) Comparison of pollen spectra collected by four different subspecies of the honey bee *Apis mellifera*. *Apidologie* **38**, 341-353.

Kovac, H., Stabentheiner, A. & Brodschneider, R. (2009) Contribution of honeybee drones of different age to colonial thermoregulation *Apidologie* **40**, 82–95

Kraus, B. & Velthuis, H.H.W. (1997) High Humidity in the Honey Bee (*Apis mellifera* L.) Brood Nest Limits Reproduction of the Parasitic Mite *Varroa jacobsoni* Oud. *Naturwissenschaften* **84**, 217-218.

Kraus, B. & Velthuis, H.H. (2001) The impact of Temperature Gradients in the Brood Nest of Honeybees on Reproduction of Varroa jacobsoni In: *Proceedings of the Second International Conference on Africanized Honey Bees and Bee Mites* (E. Erickson & R. Page, Eds). A.I. Root, Inc. 235-250.

Kraus, F. B., Neumann, P., Scharpenberg, H., Van Praagh, J. & Moritz, R. F. A. (2003) Male fitness of honeybee colonies (*Apis mellifera* L.). *J . Evol . Biol.* **16**, 914-920.

Langstroth L.L. (1866) *A practical treatise on the hive and the honey bee*, 3rd ed., Lippincott and Co., Philadephia.

Langstroth, L. L. (1853) *Langstroth on the hive and the honey bee - A Bee Keeper's Manual*, Hopkins, Bridgeman & Co. Northampton.

Le Conte, Y., de Vaublanc, G., Crauser, D., Jeanne, F., Rousselle, J-C., Bécard, J-M (2007) Honey bee colonies that have survived Varroa destructor. *Apidologie* **38**, 1-7.

Leopoldino, M. N., Freitas, B. M., Sousa, R. M. & Paulino, F. D. G. (2002) Avaliação do uso do feromônio de Nasonov sintético e óleo essencial de capim santo (Cymbopogon citratus) como atrativos para enxames de abelhas africanizadas (*Apis mellifera*). *Ciência Animal. Fortaleza* **12**(1), 19-23.

Leroy, R. (1946) *Les Abeilles et la Ruche Mixte.* Yvette Leroy. Summarised at: http://warre.biobees.com/leroy_combination_hive.pdf

Loper, G.M., Sammataro, D., Finley, J., & Cole, J. (2006) Feral Honey Bees in Southern Arizona, 10 Years After Varroa Infestation. *Am. Bee J.* **146**(6), 521-524.

Lodesani M., Costa C., Serra G., Colombo R. & Sabatini A.G. (2008) Acaricide residues in beeswax after conversion to organic beekeeping method. *Apidologie* **39**, 324-333.

Mancke, G. & Csarnietzki, P. (2005) *Der Weißenseifener Hängekorb*. Weißenseifen-Micheelshag: Werkgemeinschaft Kunst und Heilpedagogik. See also www.themelissagarden.com/beekeeping.html.

Manley, R. O. B. (1946) *Honey Farming*. Faber & Faber, London.

Manning, R.(2001) Fatty acids in pollen: a review of their importance for honey bees. *Bee World* **82**(2), 60-75.

Marris, G. (2010) Beebase - a free online resource for beekeepers. *Gwenynwyr Cymru* Number **168**, 24-27.

Mattila, H. R. & Seeley, T. D. (2007) Genetic Diversity in Honey Bee Colonies Enhances Productivity and Fitness. *Science* **362**, 317. DOI: 10.1126/science.1143046.

Mavrofridis, G. (2009) Professional beekeeping with improved traditional moveable-comb hives. *Buzz Extra* **6**(2), 4-5.

McMullan, J. B. & Brown, M. J. F. (2005) Brood pupation temperature affects the susceptibility of honeybees (Apis mellifera) to infestation by tracheal mites (*Acarapis woodi*). *Apidologie* **36**, 97-105.

Meixner, M. D., Costa, C., Kryger, P., Hatjina, F., Bouga, M., Ivanova, E. & Büchler, R. (2010) Conserving diversity and vitality for honey bee breeding. *J. Apicultural Res.* **49**(1) 85-92.

Melathopoulos, A. (2006) Honey is the Sustainable and Ethical Sweetener. *Beekeepers Quarterly* **84**, 10-11.

Mepham, B. (1996) Chapter 7: Ethical analysis of food biotechnologies: an evaluative framework. In: *Food Ethics* Mepham, B. (ed.) Routledge, London & New York. 101-119. Mepham's ethical matrix is also presented in detail in *Farming animals for food - Towards a moral menu*. Food Ethics Council Report 3. 2001. www.foodethicscouncil.org; PDF at http://www.dheaf.plus.com/warrebeekeeping/farming_animals_for_food_towards_a_moral_menu_fec_2001.pdf.

Mullin, C. A., Frazier, M., Frazier, J.L., Ashcraft, S., Simonds, R., van Engelsdorp, D. & Pettis, J. S. (2010) High Levels of Miticides and Agrochemicals in North American Apiaries: Implications for Honey Bee Health. *PLoS ONE* 5(3): e9754. doi:10.1371/journal.pone.0009754. http://www.plosone.org/article/info%3Adoi%2F10.1371%2Fjournal.pone.0009754.

Oldroyd, B. P. & Fewell, J. H. (2007) Genetic diversity promotes homeostasis in insect colonies. *Trends in Ecol. Evol.* **22**(8), 408-413.

Palmer, K. A. & Olroyd, B. P. (2003) Evidence for intra-colonial genetic variance in resistance to American foulbrood of honey bees (*Apis mellifera*) *Naturwissenschaften* **90**, 265-268.

Paton, D. C. (1996) *Overview of feral and managed honeybees in Australia: distribution, abundance, extent of interactions with native biota, evidence of impacts and future research*. Australian Nature

Conservation Agency, Canberra.

Pettigrew, A. (1875) *The Handy Book of Bees - Being a Practical Treatise on their Profitable Management.* William Blackwood and Sons, Edinburgh & London.

Piccirillo, G. A. & De Jong, D. (2003) The influence of brood comb cell size on the reproductive behavior of the ectoparasitic mite Varroa destructor in Africanized honey bee colonies. *Genet. Mol. Res.* 2(1), 36-42. www.funpecrp.com.br/gmr/year2003/vol1-2/gmr0057_full_text.htm.

Ratnieks, F.L.W. (2007) How far do honeybees forage? *Beekeepers Quarterly* **89**, 26-28.

Richard F-J., Tarpy D.R. & Grozinger C.M. (2007) Effects of Insemination Quantity on Honey Bee Queen Physiology. *Public Library of Science ONE* 2(10); www.plosone.org/doi/pone.0000980.

Rinderer, T.E., DeGuzman, L.I., Delatte, G.T. & Harper, C. (2003) An Evaluation of ARS Russian Honey Bees in Combination with other Methods for the Control of Varroa Mites. *American Bee Journal* **143**(5), 410-413.

Rinderer, T.E., Harris, J.W., Hunt, G.J. & deGuzman, L.I. (2010) Breeding for resistance to Varroa destructor in North America. *Apidologie* **41**, 409-424.

Ritter, W. (2007) Bee Death in the USA: Is the Honeybee in Danger? *Beekeepers Quarterly* **89**, 24-25.

Sammataro, D., A. Vásquez, & Olofsson, T.C. (2009) Preliminary study on the effect of artificial diet on the lactic acid bacterial flora in the honey stomach of honeybees. (in preparation).

Sasaki, M., Nakamura, J., Tani, M. & Sakai, T. (1990) Nest temperature and winter survival of a feral colony of the honeybee, Apis mellifera L., nesting in an exposed site in Japan. *Bull. Fac. Agr. Tamagawa Univ.* 30, 9-19.

Seeley, T. D. & Morse, R. A. (1976) The nest of the honey bee (*Apis mellifera L.*) *Insectes Sociaux* **23**(4), 495-512.

Seeley, T. D. & Morse, R. A. (1977) Dispersal behaviour of honey bee swarms. *Psyche* **84**(3-4), 199-209.

Seeley, T. D. & Morse, R. A. (1978) Nest site selection by the honey bee, *Apis mellifera. Insectes Sociaux* **25**(4), 323-337.

Seeley, T. D. & Tarpy, D. R. (2007) Queen promiscuity lowers disease within honeybee colonies *Proc. R. Soc. B* **274**, 67-72; doi:10.1098/rspb.2006.3702, published online 26 September 2006.

Seeley, T. D. (2002) The effect of drone comb on a honey bee colonyís production of honey *Apidologie* **33**, 75-86.

Seeley, T.D. (2007) Honey bees of the Arnot Forest: a population of feral colonies persisting with Varroa destructor in the northeastern United States. *Apidologie* **38**, 19-29.

Simone, M., Evans, J. D. & Spivak, M. (2009) Resin collection and social immunity in honey bees. *Evolution* **63**(11), 3016-3022.

Sims, D. (1997) *Sixty years with bees*. Northern Bee Books.

Singer, H. (2007) Ein möglicher Ausweg aus der Varroakrise? *Bienen Aktuell* 04/2007, Austria.

Somerville, D. (2005) *Fat bees, skinny bees - a manual on honey bee nutrition for beekeepers*. Australian Rural Industries Research and Development Corporation Publication No. 05/054.

Southwick, E. E. (1985) Thermal conductivity of wax comb and its effect on heat balance in colonial honey bees (*Apis mellifera* L.). *Experientia* **41** 1486-1487.

Spivak, M. & Reuter, G.S. (2008) New direction for the Minnesota Hygienic line of bees. *Am. Bee J.* **148**, 1085-1086.

Steiner, R. (1933) *Nine lectures on bees*. Part of GA351, Lecture V1, Dornach 10 December 1923. Transl. Pease, M. & Mirbt, C. A., Anthroposophical Agricultural Foundation.

Storch, H. (1985) *At the Hive Entrance. Observation Handbook: How to Know what happens inside the hive by observation of the outside*. Transl. from *Am Flugloch*, European Apicultural Editions, Brussels.

Strange, J. P., Arnold, G., Garnery, L. & Sheppard, W. S. (2009) Conservation of a locally adapted population of *Apis mellifera* L. An integrative approach. Proceedings of Apimondia 2009.

The Bee-friendly Beekeeper

Summers, B. (2008) 'The Summers' coffin hive. *The Beekeepers Quarterly* **93**, 19-21.

Tan, K., Bock, F., Fuchs, S., Streit, S., Brockmann, A. & Tautz, J. (2005) Effects of brood temperature on honey bee *Apis mellifera* wing morphology. *Acta. Zoologica. Sinica.* **51**, 768-771.

Tarpy, D. (2003) Genetic diversity within honeybee colonies prevents severe infections and promotes colony growth. *Proc. R. Soc. Lond. B* **270**, 99-103.

Tarpy, D. R. & Seeley, T. D. (2006) Lower disease infections in honeybee (*Apis mellifera*) colonies headed by polyandrous vs monandrous queens. *Naturwissenschaften* **93**, 195-199.

Tautz, J. (2008) *The Buzz About Bees - Biology of a Superorganism* (Springer Verlag).

Tautz, J., Maier, S., Groh, C., Rössler, W. & Brockmann, A. (2003) Behavioral performance in adult honey bees is influenced by the temperature experienced during their pupal development. *PNAS* **100**(12) 7343-7347.

Thun, M. K. (1986) *Die Biene Haltung und Pflege Unter Berücksichtigung Kosmischer Rhythmen* M. Thun-Verlag, Biedenkopf.

Thür, J. (1946) *Bienenzucht: Naturgerecht einfach und erfolgsicher*. 2nd ed. Friedrich Stock's Nachf. Karl Stropek (Buchhandlung und Antiquariat), Vienna. Chapters 1 & 2 translated as *Beekeeping: natural, simple and successful.* : Heaf, D. J. (2007); free download: www.users.callnetuk.com/~heaf/thur.pdf

Vásaquez, A. & Olofsson, T. (2009) Lactic acid bacteria: can honey bees survive without them? Proceedings of Apimondia 2009. www.apimondia.org/2009/proceedings.htm.

Veuille, G. *La ruche ronde divisible*. Gilbert Veuille, 28 Rue Bois, entrée 5, 3700 Tours, France. Tel 02 47 64 01 29. Some details of construction and use given in http://warre.biobees.com/round_warre_hives.pdf.

Villa, J. D., Bustamante, D. M., Dunkley, J. P. & Escobar, L. A. (2008) Changes in Honey Bee (Hymenoptera: Apidae) Colony Swarming and Survival Pre- and Postarrival of Varroa destructor. *Ann. Entomol. Soc. Am.* **101**(5), 867-871.

Villa, J. D., Rinderer, T. E. & Bigalk, M. (2009) Overwintering of Russian honey bees in northeastern Iowa. *Science of Bee Culture* **1**(2), 19-21.

Visscher P.K. & Seeley T.D. (1982) Foraging strategy of honeybee colonies in a temperate deciduous forest. *Ecology* **63**, 1790-1801.

Waite, R., Brown, M., Thompson, H. & Bew, M. (2003) Control of American foulbrood by eradication of infected colonies. *Apiacta* **38**, 134-136.

Warré, E. (1923) *L'Apiculture Pour Tous*, 5th edition, Tours. http://warre.biobees.com/warre_5th_edition.pdf. Pages 60-71 translated by David Heaf: http://warre.biobees.com/warre_5ed_60-71.pdf.. See also JPEG images of scans of the pages of the book at http://ruche.populaire.free.fr/apiculture_pour_tous_5eme_edition/.

Warré, É. (1948) *L' Apiculture pour Tous* 12th edition. (Saint-Symphorien). Free download of the PDF: http://www.apiculture-warre.fr/livre_warre.html. For JPEG scans of the original pages: http://ruche.populaire.free.fr/apiculture_pour_tous_12eme_edition/.

Warré, É. (2010) *Beekeeping for All*. Northern Bee Books. Transl. by Heaf, D. J. & Heaf, P. A. from *L' Apiculture pour Tous* 12th edition. (Saint-Symphorien, 1948). *Beekeeping for All* is also available as a free download at http://www.users.callnetuk.com/~heaf/beekeeping_for_all.pdf.

Webster, K. (2008) A new paradigm for American Beekeeping. *Am Bee J.* **148**(3), 257-259.

Weiler, M. (2008) Biodynamic beekeeping: an interview with Michael Weiler. *Beekeepers Quarterly* **91**, 16-18.

Winston, M. L. (1987) *The biology of the honey bee*. Harvard University Press, Cambridge, Mass.

Wray, M.K., Mattila, H.R. & Seeley, T. D. (2011) Collective personalities in honeybee colonies are linked to colony fitness. *Animal Behaviour* **81**(3) 559-568.

Wright, W. (2006) Advantages/Disadvantages of Swarm Prevention By Checkerboarding/Nectar Management, *Bee Culture* May.

Notes

1 Russian Hollow Log Beekeeping Traditions: http://www.thehoneygatherers.com/html/photolibrary16.html
2 Thun (1986)
3 Mancke & Csarnietzki (2005)
4 Hauk (2002)
5 Chandler(2007)
6 Berrevoets (2009)
7 Warré, É. (2010)
8 Conrad (2007)
9 The organisation dedicated to this in the UK is the Bee Improvement and Bee Breeders' Association: www.bibba.com.
10 Bees for Development: www.beesfordevelopment.org
11 Royal Hawaiian Honey, royalhawaiianhoney.com
12 Commission Regulation (EC) No 889/2008 of 5 September 2008. Official Journal of the European Union, 18 September 2008.
13 www.foodethicscouncil.org
14 Kockelkoren (1995)
15 Manley (1946)
16 Conrad (2007) p. 37.
17 Conrad (2007) p. 65.
18 Kockelkoren (1995)
19 Mepham (1996)
20 Mullin et al. (2010)
21 Langstroth (1853) p. 118.
22 Pettigrew (1875) p. 52.
23 For a striking example see Sasaki et al. (1990)
24 Tautz (2008)
25 Bujok et al. (2002)
26 Bujok (2005)
27 Tan et al. (2005)
28 Tautz et al. (2003)
29 Becher et al. (2009)
30 Groh et al. (2004)
31 Jones et al. (2005)
32 McMullan et al. (2005)
33 For examples, Phil Chandler: www.biobees.com; Dennis Murrell: http://beenatural.wordpress.com/top-bar-hives/
34 Warré (2010) For Warré hive modifications see also: http://warre.biobees.com/index.html.
35 Dietrich Vageler, personal communication.
36 Thür (1946)
37 Christ (1783)
38 See also Conrad (2007) for natural organic beekeeping based on Langstroth hives.
39 Cheshire (1888) p. 60.
40 Human (2006)
41 Villa (2009)
42 Seeley & Morse (1976)
43 Links to videos of Georg Klindworth's Lüneberg skep beekeeping are available at http://warre.biobees.com/pressing.htm
44 Hauk (2002)
45 Mancke & Czarnietzki (2005)
46 Thun (1986) See also www.biobees.com/forum/viewtopic.php?t=125 and http://web.utanet.at/huttinge/projekte/nepal/book_off/trainbook.htm#strawhive.
47 Mavrofridis (2009)
48 Summers (2008)
49 www.beesource.com/pov/usda/thermology/techbulletin1429.htm
50 Christ (1783)
51 Warré (2010) p. 41
52 Warré (2010) p.40
53 Langstroth (1853) p. 121.
54 Harbo & Harris (2004)
55 Rinderer et al. (2003)
56 Villa et al. (2009)
57 Human et al. (2006)
58 Kraus & Velthuis (1997)
59 Kraus & Velthuis (2001)
60 Chapleau (2003)
61 Seeley & Morse (1976)
62 Seeley (2002)
63 Langstroth (1866) p. 51.
64 Seeley & Morse (1976)
65 Richard et al. (2007)
66 Tarpy & Seeley (2006)
67 Seeley & Tarpy (2007)
68 Tarpy (2003)
69 Palmer & Olroyd (2003)
70 Jones et al. (2004)
71 Graham et al. (2006)
72 Mattila & Seeley (2007)
73 Oldroyd & Fewell (2007)
74 Kraus et al. (2003)

75   Dave Cushman http://www.dave-cushman.net/bee/cellsize.html

76   Ibid.

77   http://www.dheaf.plus.com/warrebeekeeping/cell_size.htm

78   For example: www.beesource.com/point-of-view/ed-dee-lusby/

79   Piccirillo & De Jong (2003)

80   Singer (2007)

81   Ellis et al. (2009)

82   Berry et al. (2010)

83   Coffey et al. (2010)

84   Seeley, T.D. Summary findings prior to publication in *Apidologie* at http://www.reeis.usda.gov/web/crisprojectpages/211868.html

85   Dennis Murrell http://beenatural.wordpress.com/observations/natural-comb/

86   Warré (2010) p.40

87   Berry (2009)

88   Mullin et al. (2010)

89   Michael Bush http://www.bushfarms.com/beesframewidth.htm

90   De Jong (1982)

91   Seeley & Morse (1976)

92   Berry & Delaplane (2001)

93   Manley (1946) p. 145-146.

94   Free & Williams (1974)

95   Murrell, D. Nest Structure: http://beenatural.wordpress.com/observations/nest/

96   Cushman, D. Preserving the integrity of honey bee nest structure. http://www.dave-cushman.net/bee/nest_integrity.html

97   Tautz (2008) p. 187.

98   Gillette (1900)

99   http://www.beesource.com/resources/usda/the-thermology-of-wintering-honey-bee-colonies/

100   Tautz (2008), pp. 179 & 217.

101   Links to videos of Georg Klindworth's Lüneberg skep beekeeping are available at http://warre.biobees.com/pressing.htm

102   http://www.imkerei-klindworth.de/ For a 2.5 hour streamed or purchasable film of Georg Klindworth's operation see http://tinyurl.com/y9276f9 or search 'Korbimkerei' at http://www.iwf.de.

103   Pettigrew (1875) pp. 84-86.

104   Storch (1985)

105   Precursors of the People's Hive of Abbé Emile Warré, http://warre.biobees.com/precursors.htm

106   Frèrès & Guillaume (1997) p. 45

107   Beebase, Food and Environment Research Agency, UK. Number of honey bee colonies inspected in England and Wales by MAFF and DEFRA appointed bee inspectors between 1952-2004, and recording incidence of foul brood disease over the same period. https://secure.fera.defra.gov.uk/beebase/

108   Vásaquez & Olofsson (2009)

109   Gillard (2008)

110   Waite et al. (2003)

111   Imperial College Consultants Limited (2008)

112   Marris (2010)

113   Gillard (2008)

114   Cheshire (1888)

115   http://warre.biobees.com/delon.htm

116   Bees for Development: http://www.beesfordevelopment.org

117   Visscher & Seeley (1982)

118   Ratnieks (2007)

119   Keller et al. (2005)

120   Alaux et al. (2010)

121   Ritter (2007)

122   Manning (2001)

123   Somerville (2005)

124   Köpler et al. (2007)

125   Steiner (1933) p. 55.

126   Bee Improvement and Bee Breeders' Association of Britain and Ireland: http://www.bibba.com

127   Baum et al. (2005)

128   Seeley (2007)

129   Paton (1996) p. 7.

130   Jaffé et al. (2007)

131   Bailey & Perry (2001)

132   Warré (2010) p.31

133   Frèrès & Guillaume (1997) Contains a 14-page cost-benefit anlysis of two hive types.

134   Forsgren et al. (2009)

135   Sammataro et al. (2009)

136	Vásaquez & Olofsson (2009)	172	Grozinger (2010)
137	Melathopoulos (2006)	173	Kocher et al. (2009)
138	Standards for beekeeping and hive products - June 2007. Biodynamic Agricultural Association. http://www.biodynamic.org.uk/fileadmin/user_upload/Documents/Demeter_Standards/Demeter_International_Bee_Standards.pdf	174	Richard et al. (2007)
		175	Meixner et al. (2010)
		176	A UK based organisation: www.bibba.com.
		177	Winston (1987) p. 189.
139	Conrad (2007) pp. 91-196.	178	Cushman, D. Honey bee colony assessment criteria. http://www.dave-cushman.net/bee/assessmentcriteria.html
140	Macdonald, P. Hansard written answers. Fruit trees (arsenical sprays) 20 January 1942, vol 377, c212W.	179	Pettigrew (1875) pp. 126-131.
		180	Chandler, P. (2008) The Barefoot Beekeeper: http://www.biobees.com
141	Berry (2009)	181	http://www.ruche-warre.com
142	Mullin et al. (2010)	182	http://warre.biobees.com/denis.htm
143	Weiler (2008)	183	Seeley & Morse (1978)
144	Atienza et al. (1993)	184	Seeley & Morse (1977)
145	Ellis et al. (2009b)	185	Leopoldino et al. (2002)
146	Le Conte et al. (2007)	186	Berrevoets (2009) p. 101.
147	Fries et al. (2006)	187	Warré (2010) pp. 35 & 37.
148	Fries & Bommarco (2007)	188	Imperial College Consultants Limited (2008) p. 5
149	Villa et al. (2008)		
150	Loper et al. (2006)	189	Seeley & Morse (1976)
151	Seeley (2007)	190	Wright (2006)
152	Villa et al. (2008)	191	Seeley & Morse (1976)
153	Alphonse, J. (2009) Personal communication. http://www.mielleriealphonse.com/index.html	192	Christ (1783)
		193	Warré (2010). L'Abbé Eloi François Émile Warré was born on 9 March 1867 at Grébault-Mesnil in the Somme département. He was ordained a priest on 19 September 1891 - Amiens diocese - and became the parish priest of Mérélessart (Somme) in 1897 then of Martainneville (Somme) in 1904. He disappeared from the records in 1916 subsequently to reappear at Saint-Symphorien (Indre-et-Loire) to devote himself exclusively to beekeeping. He died at Tours on 20 April 1951. Abbé Warré developed The People's Hive based on his studies of 350 hives of different systems that existed at his time as well as of the natural habits of the bee. To publish his findings he wrote several books: *La santé ou les Meilleurs traitements de toutes les maladies* (Health or better treatments for all illnesses), *Le Miel, ses propriétés et ses usages* (Honey its properties and applications), *La Santé, manuel-guide des*
154	Dewhurst (2006)		
155	Webster (2008)		
156	Vásaquez & Olofsson (2009)		
157	Goodwin et al. (1994)		
158	Gilliam (1997)		
159	Evans (2006)		
160	Simone et al. (2009)		
161	Vásaquez & Olofsson (2009)		
162	Fries & Camazine (2001)		
163	Goodwin et al. (1994)		
164	Frazier et al. (2008); Mullin et al. (2010)		
165	Alaux et al. (2009)		
166	Ratnieks (2007)		
167	Imperial College Consultants Limited (2008)		
168	Steiner (1933)		
169	Imperial College Consultants Limited (2008)		
170	Warré (2010) p. 150.		
171	Keeling et al. (2003)		

malades et des bien-portants (Health, a manual for the ill and the well) - 1912 - and by far the most important *L'Apiculture pour Tous* (Beekeeping for All) whose twelfth and last edition is dated 1948.

194 Warré (1948)

195 http://uk.groups.yahoo.com/group/ warrebeekeeping.

196 Links to various language Warré e-groups, fora and web sites care at http://warre. biobees.com/links.htm.

197 Warré (2010) p. 147.

198 Precursors of the People's Hive of Abbé Emile Warré: http://warre.biobees.com/ precursors.htm.

199 Pettigrew (1875) p. 85.

200 A list of Warré hive suppliers is available on the following web page: http://warre. biobees.com/links.htm

201 The Brittish Beekeepers' Association keeps lists of local associations: http:// www.britishbee.org.uk.

202 Waxing top-bars: www.dheaf.plus.com/ warrebeekeeping/waxing_topbars.htm

203 Frèrès & Guillaume (1997) p. 46.

204 Preparing hessian: http://www.dheaf.plus. com/warrebeekeeping/preparing_hessian. htm

205 http://warre.biobees.com/plans.htm

206 Hampshire, N. *Warré Hive Construction Guide.* http://thebeespace. net/2008/07/30/introduction-warre-beehive-construction-guide/.

207 Warré (2010) p. 83ff

208 Comb transfer to a Warré hive: http:// warre.biobees.com/comb_transfer.htm

209 Warré (2010) pp. 60 & 62.

210 Storch (1985)

211 http://warre.biobees.com/pressing.htm

212 Pressing honey out of crushed comb: http://www.dheaf.plus.com/ warrebeekeeping/pressing_honey.htm

213 Sims (1997) p. 102.

214 Solar extractor: www.dheaf.plus.com/ warrebeekeeping/solar_extractor.htm

215 Warré (2010) p. 98.

216 Murrell, D. Varroa Blaster - Sugar dusting mites into oblivion. http://beenatural. wordpress.com/stuff/varroa-blaster/

217 Warré (2010) p. 54.

218 Ibid. p. 26.

219 Warré beekeeping e-group: http:// uk.groups.yahoo.com/group/ warrebeekeeping

220 Harbo & Harris (2004)

221 Seeley & Morse (1976)

222 Ibid.

223 Ibid.

224 Frèrè s & Guillaume (1997) p. 45

225 Gatineau (2007) See also Dav Croteau's hive with the holes fitted with plastic shutters: http://warre.biobees.com/ croteau.htm.

226 Seeley & Morse (1976)

227 Gedde (1677)

228 Dardenne, J-F. *The pleasure of beekeeping with the Warré hive.* Downloadable as a PDF at http://warre. biobees.com/round_and_polygonal.htm.

229 Warré (2010) p. 40.

230 Veuille p. 23

231 Warré (2010) p. 51

232 Gedde (1677)

233 Veuille p. 28

234 Morimoto, S. http://warre.biobees.com/ japan.htm

235 Warré (1923) pp. 60-71

236 Gatineau (2007)

237 Denis (2008)

238 For a comparison of frames used see also: http://warre.biobees.com/frames. htm

239 Warré (1923) pp. 60-71

240 Dardenne, J-F. *Stable-Climate Hive - 'Nature's Method'* Translated by David Heaf: http://warre.biobees.com/delon. htm.

241 Denis (2008)

242 Gillard (2008)

243 Frèrès & Guillaume (1997) p. 52

244 Warré (2010) p. 53.

245 Frèrès & Guillaume (1997) pp. 243-244.

246 Frèrès & Guillaume (1997) pp. 50-53 For diagrams see http://warre.biobees.com/ guillaume.htm.

247 Warré (2010) pp. 46-48.

248 Southwick (1985)

249 http://warre.biobees.com/lift.htm

ADDENDUM

p. 15ff Fundamental attitudes of beekeepers

A number of conventional beekeepers have found my identification of fundamental attitudes of beekeepers somewhat offensive, and have interpreted these attitudes as characterising certain types of people. It should be noted that not only are the four attitudes presented equally justifiable ethically, but also one and the same individual may shift from one attitude to another depending on the task in hand, their circumstances and/or their degree of experience. As people are individuals, i.e. unique, and thus cannot reliably be classified into types, the categories presented are merely attitudes that can be freely adopted or not as a given individual decides.

p. 24 Retention of nest scent and heat

Whilst it is true that bees can compensate for increased heat loss ranging from that caused by the chimney effect of the two situations just mentioned, to complete exposure to the weather in the case of a nest hanging in tree branches, the beekeeper may wish to minimise unnecessary loss of heat, as that comes at a cost in terms of the energy, and thus nectar requirements of the colony. This is of course always to be balanced by a hive design that puts cooling the hive in hot weather within the bee's capability.

p. 25 Warrés in tropical climates

The only modification for the tropics that has come to my attention so far is to increase the possibilities for cooling the hive. One user has included a mesh floor and removed the quilt altogether.

p. 32 Hybrid straw-wood circular hive

Leroy's (1946) 'mixed hive' of straw and wood overcomes the problem of achieving circularity by using straw sandwiched between two sheets of 'unrolled' poplar.

p. 40 Role of drones

A recent detailed analysis of heat production by drones showed that they contribute significantly to colony thermoregulation (Kovac *et al.* 2009). Drones

produce about 1.5 times the amount of heat produced by a worker bee
(Harrison 1987).

p. 41 Small cells and Varroa

For a review of all published work on the effect of small cells on Varroa
reproduction up to early 2011 see Heaf (2011).

p. 42 Pesticides in foundation

Three years into a conversion to organic involving 50% annual replacement of
combs with new foundation, new combs were still showing the presence of
acaricides, particularly fluvalinate (Lodesani et al. 2008). This study illustrates
how persistent the wax-soluble pesticides can be.

p. 43 Sensitivity of bees to magnetic fields

Likely receptors for magnetic fields are the superparamagnetic particles found
in cells just under the cuticle of the bee abdomen (Hsu et al. 2007). Exposure
of these particles to magnetic fields causes size changes and corresponding
intracellular electrolyte flows which could trigger neural pathways. Research on
bee sensitivity to magnetic fields is briefly reviewed in the same paper.

p. 58 Adaptation to local flora

A recent study in Les Landes area of France has identified an ecotype of *Apis
mellifera* that has adapted its annual brood cycle to the seasonal changes in the
local flora (Strange *et al.* 2009).

p. 63 Toxicity of thymol

In an extensive review of the toxicity of pesticides to honey bees, the authors
warned that thymol may harm bees, despite its natural origins (Johnson *et al.*
2010).

p. 65 Breeding Varroa tolerant bees

Initiatives to produce Varroa tolerant bees have found solutions with varying
degrees of success in several countries: France – John Kefuss (Kefuss *et al.*
2009); UK – Ron Hoskins; USA – Marla Spivak (Spivak & Reuter 2008) and

Thomas Rinderer (Rinderer *et al.* 2010). Increasingly, commercial breeders, for example Kirk Webster, have built on these successes, and there are also many 'non treaters' who have bred from their own local survivors of Varroa and achieved winter survival rates and longevities of colonies that they find tolerable. Both Kefuss and Webster emphasise the approach that has long been a principle of organic livestock breeding, namely that it should be local and integrated with the way the livestock are managed.

p. 74-75 Trait selection versus holistic selection during breeding

One trait commonly selected for during breeding is docility. However, Wray *et al.* (2011) have shown in a study on colony personality that 'the more defensive colonies will subsequently go on to be more productive, grow larger, and have a higher likelihood of winter survival'. This finding, also noticed by others, warns us that overly focusing on individual traits might have to be paid for by a reduction in colony fitness.

p. 101 Warré colony losses

In the exceptionally hard winter of 2010/2011, my colony loss rate doubled to 60%. The causes were the unusual continuation of brood rearing into late autumn of 2010 due to the extended and productive nectar flows from Himalayan balsam and ivy, the reduced colony populations in autumn as a result of the pressure from Varroa, the consequent reduced number of winter bees, and finally the severe cold spells in November and December 2010 with temperatures down to minus eleven degrees Celsius. Emerging brood had died in their cells in most colonies that had failed, and small clusters of adult bees had chilled next to copious stores of honey. However, at the time of writing this addendum, the colony illustrated in Fig. 9.24 is in its 5th season and on warm days shows the same busy entrance traffic as shown in the figure.

p. 114 Vertical orientation of top-bars

It is still too early to report how colonies in Warrés, standard or modified, behave with vertically orientated bars. Two experiments have shown that the bees have no problems fixing combs to them. The sides of the bars tend to have most of the attached comb towards the bottom. So far, any cells there have been filled with honey. However, it would be of particular interest to know how the comb is continued between one box and the next. When this transition coincides with the brood area, it is highly unlikely that brood cells will be constructed against the sides of the bars. This could harm brood nest integrity more than having horizontal bars. There are no reports yet of Warré colonies that have moved down into a second box with vertically oriented bars.

p. 122 Removing combs on top-bars

The author finds it helpful to use a simple stand for supporting combs on top-bars when they are removed from the hive.

Warré comb being inspected by a government bee inspector.
Photo: John Haverson.

p. 122 Intensely melliferous localities

The Alberta beekeeper in question had four out of seven hives survive the 2010/2011 winter, all in good shape and with reserves of honey. In that climate the bees are confined for six months of the year and snowfalls can occur as late as June. His hives now have walls that are two inches (50 mm) thick and the internal dimensions of the boxes have been increased to 13 x 15 inches (330 x 380 mm). Instead of the Warré quilt, a top feeder is in place all the year round, filled with insulation when not used for feeding. The feeder has built into it a small top entrance to the outside. The bees, imported from New Zealand, are wintered on three boxes to ensure sufficient stores for the long winter.

Index

_____ 153